Basic Laboratory Skills

NIPA® GENX ELECTRONIC RESOURCES & SOLUTIONS P. LTD.
New Delhi-110 034

About the Authors

Dr. Suresh Kumar is Principal Scientist, USDA Norman E. Borlaug Fellow, Indo-US Science & Technology Forum (IUSSTF) Fellow, NSA Vising Fellow, and One of the Top 2% Most Influential Scientists of the World (2023) at the Division of Biochemistry, ICAR-Indian Agricultural Research Institute, New Delhi. He completed M. Sc. (Plant Biotechnology) from Centre for Plant Molecular Biology, Tamil Nadu Agricultural University, Coimbatore, and Ph. D. (Mol. Biol. & Biotech) from the Post-Graduate School, ICAR-I.A.R.I., New Delhi. After completing his Post-Doctoral research at University of Missouri, Columbia; University of California, Reverside and Purdue University, Indiana, USA, he is working on epigenomics of abiotic stress tolerance in plants as Principal Investigator. He has been involved in teaching and research guidance to M. Sc. and Ph. D. students of the Graduate School, ICAR-I.A.R.I., New Delhi. He is Course Leader for UG and PG courses, including the Basic Concepts in Laboratory Techniques (PGS 504), offered by the Graduate School, ICAR-I.A.R.I., New Delhi, India.

Dr. Simardeep Kaur is working as an ARS-Scientist (Plant Biochemistry) at ICAR-Research Complex for North Eastern Hill Region, Umiam, Meghalaya, India. She had completed her graduation in Biochemistry (2017) from Punjab Agricultural University (PAU), Ludhiana, then attended Post Graduate School PAU for her M.Sc. in Plant Biochemistry (2019), where she worked on Biochemical basis of leaf-blight resistance in barley and published her significant findings in high impact journal. During her Masters she received Gold medal for excellence in chemistry/ biochemistry, PAU Merit fellowship, Pyara Singh Parmar award, and qualified CSIR-JRF (NET). Later, she completed her Ph.D. (Plant Biochemistry) at ICAR-Indian Agricultural Research Institute (IARI), New Delhi. She is recipient of JRF and SRF from the CSIR during her Ph.D. She has also several cleared other national level examinations like GATE (Life Sciences), ICAR-NET, JNUEE etc. She also grabbed the opportunity to continue part of her Ph.D. program at the Utah State University, U.S.A. on receiving the most prestigious Nehru-Fulbright Doctoral fellowship (2022-23). She has published several research and review papers in high impact journals including Planta, Journal of Plant Growth Regulation, Current Plant Biology, Heliyon etc. She was also awarded the Springer Nature Young Scientist Award for her Ph.D. work at an international conference. Currently, she is developing NIRS-based predictive models for key nutritional traits in orphan or underutilized crops of Southeast Asia. In addition, she is investigating the effects of concurrent abiotic stress responses in indigenous rice varieties of Northeast India.

Basic Laboratory Skills
Volume 02: The Protocol Series

Suresh Kumar Ph.D., ARS
USDA Norman E Borlaug Fellow & IUSSTF Fellow
Principal Scientist (Plant Biochemistry)
Top 2% Most Influential Scientist of the World, 2023
Principal Investigator NASF & EMR-ICAR Projects
Division of Biochemistry
ICAR-I.A.R.I., New Delhi-110012, India

Simardeep Kaur Ph.D., ARS
Nehru Doctoral Fulbright Fellow
Scientist (Plant Biochemistry)
Division of Crop Science,
ICAR-Research Complex for North Eastern Hill Region (NEH)
Umiam, Meghalaya, India

NIPA® GENX ELECTRONIC RESOURCES & SOLUTIONS P. LTD.
New Delhi-110 034

NIPA® GENX ELECTRONIC RESOURCES & SOLUTIONS P. LTD.

101,103, Vikas Surya Plaza, CU Block
L.S.C.Market, Pitam Pura, New Delhi-110 034
Ph : +91 11 27341616, 27341717, 27341718
E-mail: newindiapublishingagency@gmail.com
www: www.nipabooks.com

For customer assistance, please contact
Phone: + 91-11-27 34 17 17
Fax: + 91-11- 27 34 16 16
E-Mail: feedbacks@nipabooks.com

Print ISBN: 978-93-58872-18-7

ebook ISBN: 978-93-58877-46-5

Composed and Designed by NIPA®.

त्रिलोचन महापात्र
अध्यक्ष
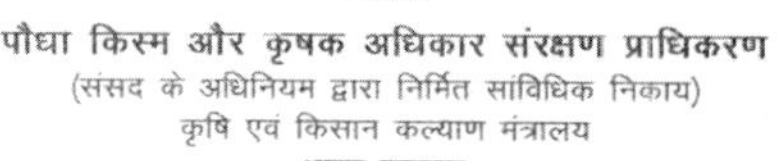

Trilochan Mohapatra
Chairperson
Protection of Plant Varieties and Farmers' Rights Authority
(A Statutory Body created by an Act of Parliament)
Ministry of Agriculture and Farmers Welfare
Government of India

Foreword

The knowledge of basic laboratory procedures is important for safety and success in research work. Skill helps avoiding error and repeated experimentation to save one's time and laboratory consumables. More importantly, personal and environmental safety during experiments is very crucial. Therefore, every researcher needs to have appropriate knowledge of basic laboratory skills, which must be acquired preferably during apprenticeship.

"Basic Laboratory Skills" written by Dr. Suresh Kumar and Dr. Simardeep Kaur is a comprehensive resource material for the students undertaking a course on Basic Concepts of Laboratory Techniques". It consists of 21 chapters covering all relevant aspects of laboratory techniques. The chapters emphasize on practical aspects of laboratory procedures, as well as instruments including safety measures to be followed while conducting research work. Objective type questions along with their answers are also included at the end of each chapter, which will help students meet their academic requirements as well as to prepare for various examinations. The chapters devoted to the usage, arrangement, storage and handling of laboratory chemicals and equipment also describe do's and don'ts for the sake of personal and environmental safety in laboratory setups. The information on personal safety measures, biosafety concerns, and ethical issues provided in simple and comprehensive language would enable the readers to easily comprehend and acquire competence. With illustrative figures and elaborated contents, this book is expected to be a valuable study material for students, technicians, researchers, professionals, and scientists in the fields of biological and chemical sciences.

Relevance of this book is evident from the increasing importance of laboratory activity in education and research, wherein necessity of following safety protocols is very important in scientific investigations. The book addresses some of the critical gaps by providing instruction/guidance often not covered in standard textbook. The book aims at enabling students/researchers to be prepared for handling laboratory chemicals and equipment safely with efficiency.

I sincerely appreciate the authors for bringing out such a updated and useful resource publication. I do hope that this valuable effort of the authors will provide the much needed reading material for students and researchers, which would support laboratory and research activities in a safe and successful manner.

Dated: 31st July, 2024

(Trilochan Mohapatra)

एन ए एस सी काम्प्लैक्स, डी पी एस मार्ग, नई दिल्ली–110012
NASC Complex, DPS Marg, New Delhi-110012
Tel: 011-25848127, 25841696; Mob.: +91-7840000542; Fax: 011-25840478
Website: www.plantauthority.gov.in
E-mail: chairperson-ppvfra@nic.in

Preface

Zero error analysis can be obtained only by zero accident, and zero laboratory accident requires hundred percent commitments to safety

This book has been designed to serve two primary goals: (i) for teaching basic concepts of laboratory techniques and (ii) following good laboratory practices while working in laboratory or doing research. The book presents the importance of laboratory skills/practices and biosafety issues to students and researchers. Some of the Do's and Don'ts provided in this books show how theoretical knowledge and practical situations are integrated in making research work successful and sustainable. With emphasis on laboratory skills, maintenance and usage of basic laboratory equipment, handling of hazardous chemicals/solutions, microbiological preparations, tissue culture techniques, biosafety, and ethical issues, the book encourages students to understand and practice safety techniques so that they can inculcate such practices in their professional life. The chapters included in this book cover information on handling of laboratory chemicals, equipment, preparation of buffers, media, basics of seed testing, biosafety, etc. which constitute curriculum of the course Basic Concepts in Laboratory Techniques (PGS504, 0L + 1P) for Post-Graduate students of the Indian National Agricultural Research System (NARS) including Deemed Universities, State Agricultural Universities (SAUs), and Central Agricultural Universities (CAUs). In the interest of the students, objective type questions along with the answers have also been included at the end of every chapter that might help preparing for various competitive examinations/ interview. Thus, we aim to provide readily available resource on basic concepts in laboratory research which are necessary for the students of SAUs, CAUs Private/Central/State institutions, clinicians, scientists, healthcare professionals, and researchers. As most of the chapters in this book have been written in simple language with brief and pertinent information for students and researchers who are looking forward to have better understanding of the basic concepts in laboratory techniques; we believe that the book will be very useful for each and every spirant/audience.

Suresh Kumar
Simardeep Kaur

Contents

1

Handling of Chemicals and Solutions in Laboratory

~Use of chemicals is like a lock, but you are the key!

Introduction

For proper and efficient experiment to be conducted, experimenters need so many chemicals to deal with. This includes different acids, base, salt, different kind of indicators and so on. Some of these chemicals are safe for health, but some of them are dangerous to health when come in close proximity to our body while some of the chemicals cause health issues on inhaling. These may cause improper functioning of some of our physiological activities. Acids and bases are very dangerous for our tissues when come in our contact. These may cause serious respiratory problem in addition to injuries on coming to our contact. Some chemicals are exploratory yet they may cause even death of the person working with them in laboratory. Therefore, mere use of the chemicals in laboratory is not only the objective of the experimenter. The person working in laboratory should also be concerned about the proper use and handing of the chemical while working with it. Thus, every researcher of an institution who has to work with chemicals in a laboratory should be made aware about the chemicals to be used in the laboratory and the consequences if any mishandling of a laboratory chemical occurs. Not only these, suitable precautionary measures should also be known to the researcher along with ensuring the availability of the required safety equipment/installations. Hence, a proper understanding of the ways of handling of chemicals in a laboratory is of prime importance for the personals like you who are working or have to work in laboratory.

Safe Handling of Chemicals in Laboratory

Safe use/handling of chemicals in laboratory include minimized exposure to chemicals, maintaining safety data sheets, proper labelling, storage, segregation, and transport of chemicals for personal hygiene.

1. Minimizing Exposure to Chemical

To minimize exposure to any chemical one must follow the below guidelines:

- As far as possible use (substitute) less hazardous chemical.
- If possible, use smaller quantity of chemical.
- Reduce the possible entry routes of chemical (such as inhalation, ingestion, skin, eye, and injection) into the body adapting standard operating procedure like use of engineering, administrative, and personal protective equipment.
- Use appropriate personal protective equipment (PPE) and other protective equipment.
- Never practice mouth pipetting.
- Avoid smelling and tasting chemicals to identify them.
- Never underestimate the potential hazards of chemicals even when substance appears to be harmless.
- Consider a substance of unknown toxicity to be toxic until proven to be safe.
- Always clean the spills created during the use of any chemical.
- When working in a cold room, keep the toxic and flammable chemicals tightly closed, as cold rooms are maintained with recirculating air.
- Never apply any beauty products/cosmetics in the areas where hazardous chemicals are stored/being used.
- Never store any eating and/or drinking item in the refrigerator where chemicals are stored.

2. Safety Data Sheet

A safety data sheet (SDS), formerly MSDS, is designed to provide emergency measure to the users of a hazardous material with the appropriate procedures for handling it. The SDS includes the following information:

- Health, physical and ecological hazards associated with the chemical/material.
- Physical properties, reactivity, and toxicological data of the chemical.
- Measures like first aid, storage, disposal, exposure control, and spill/leak procedures.

- Identity of the chemical substance.
- Physical and chemical characteristics.
- Carcinogenic and reproductive health status.
- Emergency and first aid questions.

3. Labelling of Chemical

Even if a container or reagent bottle contains a clear and safe liquid, say for example water, it must be properly labelled with the information about its content. This is the simplest rule of working in a laboratory. A chemical laboratory must be maintained in such a way that make the researchers aware about the potential risks/hazards of the chemicals in the laboratory, avoiding generation of any hazard, emergency/alternative responses like cleaning the spills and procuring appropriate medical facilities/treatment, if needed.

4. Storage of Chemicals

Centralized stockroom, storage rooms, laboratory cabinets, shelves, freezers, and refrigerators are all examples of common storage areas in research laboratories. For proper chemical storage, there are specified legal standards as well as recommended procedures. Proper storage of chemicals provides safer and healthier working environment in a laboratory, extends the life of the chemicals, as well as help avoiding contaminations.

Improper storage of chemicals may result in

- Deteriorated/damaged containers can produce dangerous fumes that may be harmful to the health of laboratory personnel
- Containers that are degraded and releasing fumes can compromise the integrity of the chemicals/containers in surrounding.
- Chemicals may become unstable and/or explosive.
- The concerned regulatory agencies may issue memorandum and/or fine.

Well-organized storage of chemicals in a laboratory helps to make better use of laboratory space and resources. Therefore, to ensure safe and durable use of chemicals, laboratories must follow certain rules. The following points should be considered while arranging chemicals in a laboratory:

- Labelling of every container for a chemical or solution is essential.

- Name of the chemical/constituent(s)/concentration as well as the associated hazard(s) should be mentioned on the label.
- The chemical containers should be checked regularly and if necessary the label should be changed before it becomes unrecognised.
- After using the chemicals, keep the containers closed (tightly closed for hygroscopic chemicals like Calcium chloride).
- Every chemical need to be stored at appropriate temperature at proper location and it must be returned to its original location after use.
- In a chemical storage cabinet, larger chemical bottles should be kept behind the smaller bottles so that smaller bottles and their content can be easily visualised.
- While storing the chemicals, the label on bottle should be facing out so that information about the chemical can be easily seen or accessed readily.
- Do not store chemicals on floor, as the container/bottle may get broken or chemicals may spill over the floor.
- If multiple containers of a chemical are present then older stock should be positioned in front of the newer chemical/stock, and the container opened or being in use should be placed in front of the unopened containers.
- If possible, the date of first opening the chemical container should be mentioned along with the name/signature of the person opening it.
- Chemicals should be kept out of direct sunlight and away from any source of heat.
- While reordering for the supply of chemicals, old stock should be checked and only that amount of chemical should be ordered which can be utilised/consumed within a given (say for example six months or a year) period of time. This helps in safe storage, avoiding any wastage (due to expiry) of chemicals, and help saving the resources/funds.
- Liquid chemical containers should be placed in secondary containment, such as trays, thermocol container to avoid breakage of bottle or leakage of the chemical.
- Chemicals should be separated and stored according to their compatibility and danger category.

- The arrival date of chemicals in the laboratory should be mentioned on the container and expired chemical should be discarded/removed from the laboratory on regular basis.
- Flammable chemicals/materials of specified flammability classes must be stored in flammable liquid storage cabinet only.
- Harmful chemicals like acids and inflammable liquids should be stored in the specified storage cabinets.
- Corrosive chemicals should be kept in separate cabinet that is resistant to corrosion.
- Organic acids like Acetic acid, Lactic acid, and Formic acid are flammable/combustible and corrosive, which should be stored in flammable or corrosive storage cabinet.
- No chemical should be stored at excess height (above the eye level), as far as possible.
- Flammable chemicals should be stored in only such refrigerator or freezer that has been approved by the manufacturer for storage of such chemicals.
- Inorganic cyanides, pyrophorics and water reactive compounds must be kept in closed storage cabinets that are kept away at safe distance from rest of the chemicals.
- Everyone in the laboratory must be appropriately instructed to use spill/ safety kit on spill of a chemical and the kit should be made available at known/informed place in the laboratory.

5. Transportation of Chemicals

The following recommendations should be followed while transporting chemicals from one laboratory another to safeguard people and the environment as well as to minimize the risk of spill.

- Always use a secondary container like a rubberized acid carrying bucket or plastic bucket while moving chemicals by hand/personally. To prevent bottles from tipping over or shattering during the transport, a small amount of appropriate packaging material such as cardboard, thermocol can be inserts between them wherever necessary.
- If possible, wheel cart with lipped surface (such as Rubberized cart) should be used for transport of glass bottles.

- Do not use vehicle or elevator having passengers for carrying chemicals. Use only freight elevators.
- In case, a passenger elevator is needed to be used for transport of chemicals, make it sure that no passengers are present in the elevator/lift.
- While carrying compressed gas cylinder, fasten the cylinder to cart and a cap should be in place on the cylinder. A compressed gas cylinder should never be thrown, rolled or dragged.

6. Segregation of Chemicals

Always store chemicals according to their compatible groups and potential health hazard, instead of storing them merely based on the alphabet of the name or by any other means. Figure 1.1 shows the compatible storage groups for proper segregation of chemicals. Storage of incompatible groups may lead to any of the following damaging/dangerous condition:

- Heat generation, fire or explosion,
- Release of toxic or flammable gas/vapour,
- Production of highly toxic or friction sensitive compound,
- Violent polymerization.

Furthermore, while segregating chemicals following points should be kept in mind:

- Physical and health hazards from the chemical,
- The form (solid, liquid or gas) of the chemicals,
- Concentration (strength) of the chemical.

Chemicals of different hazard groups (such as acids and bases) can be stored in a same cabinet as long as they are physically separated from each other, for example using high sides or deep trays. Always segregate oxidizers from flammables. Also, segregate compounds like inorganic cyanides from acids.

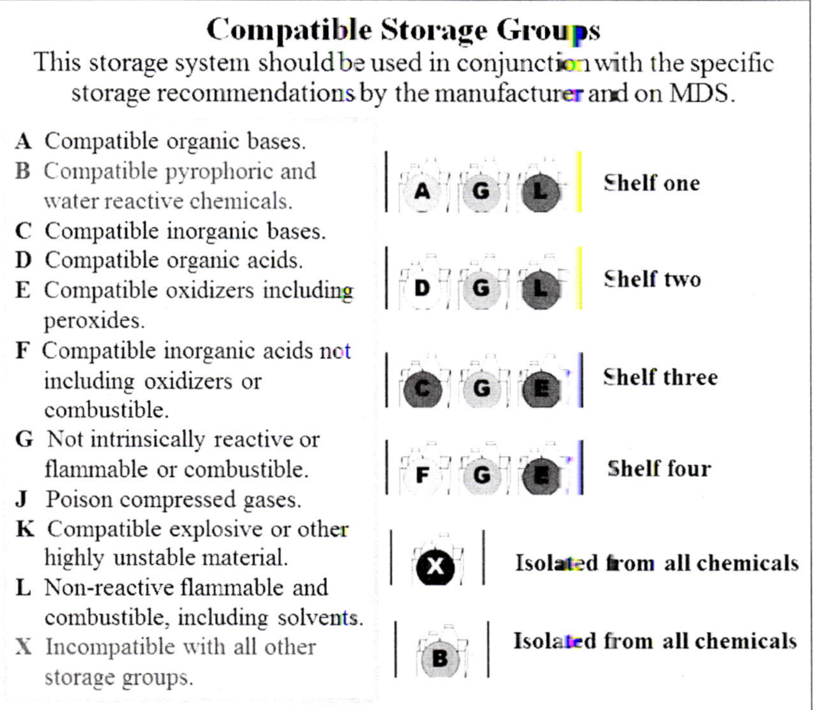

Fig. 1.1: Compatible storage group of chemicals in a laboratory

7. Personal Hygiene

For personal safety reasons, the researchers must always take care of the following points, particularly to protect themselves from the unintentional/ unnecessary use/misuse of the laboratory chemicals:

- Never store food/beverages in laboratory chemical storage area, refrigerator, or in glassware in the laboratory.
- Do not taste or smell chemicals/solutions.
- Before leaving the laboratory, remove your labcoat and wash your hands properly.
- Avoid keeping long/unbound hairs and bearing loose clothing.
- Always wear full-sleeve clothing that completely covers arms and legs.
- Wear shoes while working in a laboratory; avoid wearing sandals, perforated shoes or sneakers.

- Avoid wearing jewellery (ring, bangle, etc.) in a laboratory that interferes with use of gloves and other protective equipment. Any metallic jewellery may come in contact of electrical wiring or react with chemicals in use.

General Rules

- Always label the containers containing chemical with necessary information.
- Use the necessary personal protective equipment while handling laboratory chemicals.
- Avoid smelling, inhaling or tasting any laboratory chemical/solution.
- Never touch a chemical, keep them away from your body (hand, leg, face), clothes and shoes.
- Hazardous chemical should be handled carefully following the appropriate method for it.
- Use a separate cabinet for storing concentrated (>6 M) acids.
- Material safety data sheet (SDS) must be used to understand the nature of the chemical before handling/using a chemical. This provides many special precautionary information like storage conditions/temperature, refrigeration requirement, etc.
- A periodic check on chemical containers against rust, corrosion, and leakage should be made.
- Avoid mouth pipetting, instead use a pipette bulb or other appropriate device.
- Smoking, drinking, eating and the use of cosmetics must be avoided in laboratory/areas where hazardous chemicals are used/stored.
- For volatile chemicals, check the Standard Operating Procedure to work out what type of ventilation would be required. Always use the chemical with adequate ventilation.
- Be careful to minimize the formation of aerosol. Confine aerosols as close as possible to their source of generation, use a biosafety cabinet.

Take Home Message

While working with chemicals, keep in mind (Fig. 1.2)

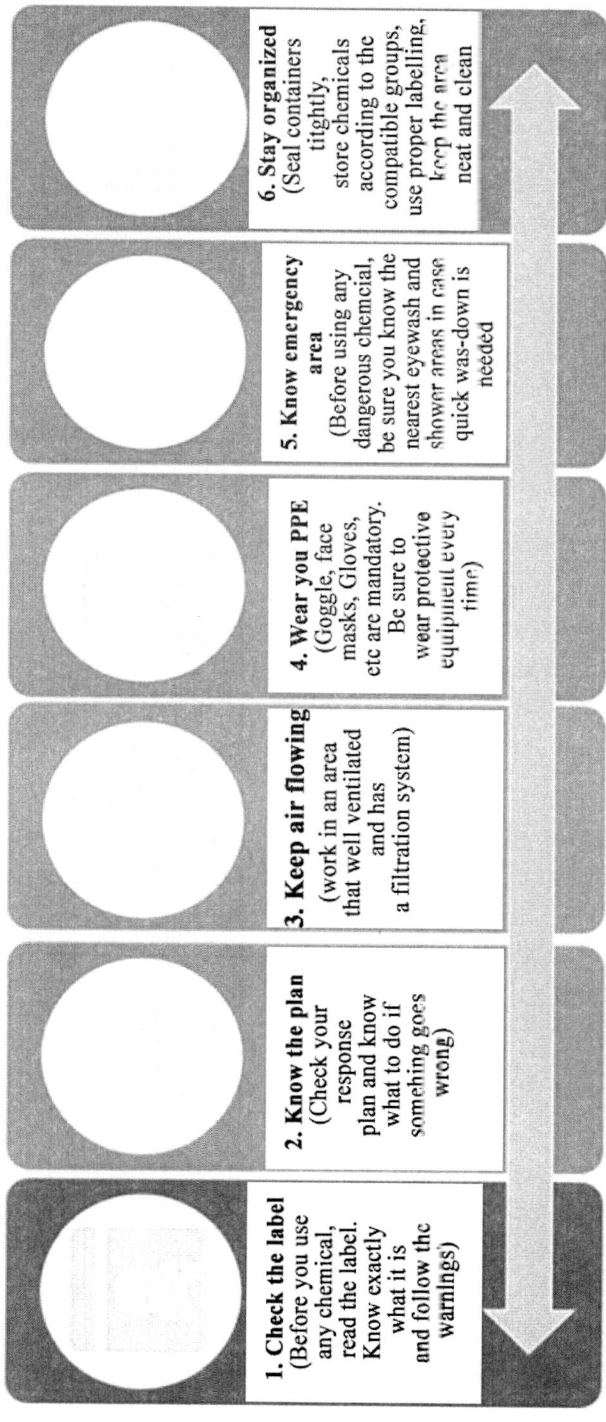

Fig.1.2: Safe handling of chemicals in laboratory

Let's Check Your Understanding

1. What do you mean by SDS in context of laboratory safety? What type of information is included in the SDS?
2. Inorganic cyanides, pyrophoric and water reactive compounds must be kept in closed storage cabinets that are situated at a safe distance from rest of the chemicals.

   True / False
3. Which cryogenic liquid is used in storing samples for a longer time?
4. What are the possible consequences of storing incompatible chemicals all together?
5. Storing food or beverages in chemical storage areas, refrigerators, glassware or else in laboratory is a good practice.

   True / False
6. What do you mean by PPE?
7. Applying beauty products/cosmetics in a laboratory, particularly in the area where hazardous chemicals are handled, will not lead to any causality in the laboratory.

   True / False

Answers

1. SDS stands for Safety Data Sheet

It must include:

- The identity of the chemical substance,
- Physical and chemical characteristics,
- Health, physical and ecological hazards associated with the material,
- Physical properties, reactivity, and toxicological data of the chemical,
- Measures like first aid, storage, disposal, exposure control, and spill/leak procedures,
- Carcinogenic and reproductive health status,
- Emergency and first aid questions.

2. True

3. Liquid nitrogen

4. Storage of incompatible group of chemicals may lead to the following conditions:

- Heat generation, fire and explosion,
- Toxic or flammable gas/vapour release,
- Synthesis of highly toxic compounds,
- Violent polymerization

5. False

6. Personal protective equipment

7. False

2

Drying Chemicals and Solutions in Laboratory

~If you're not the part of the solution, you're part of the precipitate!

Introduction

While experimenting with chemicals, sometimes we need to dry the chemical/solution (solid, liquid or gas) using a process like evaporation or distillation, mainly to remove water or different solvents. Drying is the process of evaporation or distillation to remove water/solvent from a solution, solid or even from gas. Generally, a drying agent (usually an anhydrous inorganic salt) is used for this process, which reacts with the water (present in the chemical/solvent to be dried) to form a hydrate. For example, anhydrous $MgSO_4$ is one of the drying agents. The drying agents are recognized by their capacity (the rate at which they absorb water) and their intensity (the amount of water left behind in the solvent at equilibrium). No chemical reaction or change in phase occurs during the drying process.

Type of Solvents

Apart from water (a universal solvent), there are three families of organic solvents used in laboratories.

1) Oxygenated solvents: Alcohols, Ketones, Glycol ethers, Glycol esters, etc.
2) Hydrocarbon solvents: Aliphatic and Aromatic hydrocarbons.
3) Halogenated solvents: Chlorinated hydrocarbons.

Removal of Water

Chemically, water has diverse activity. It can act as a weak acid, a weak base, a nucleophile, and sometimes in certain organometallic reactions it acts as a ligand for the metal. When the reagents used for a reaction show sensitivity to water (as in case of Grignard reagents or enolates), it gives erroneous reaction. This may create difficulties, if water is one of the products of the reaction. Therefore, use of anhydrous $CaCl_2$, Na_2SO_4, $CaSO_4$ and $MgSO_4$ as drying agent is essential. Organic liquids are considered to be wet if they contain even a fraction of water. But they remain liquid even after it is fully dried (devoid of water).

Methods of Drying

There can be two methods of drying

1. **Static drying-** In this process of drying, a drying- agents is added to the liquid solvent/chemical, allowed to stand for a while, and then mixed with a magnetic stirrer. After that it is either shaken or boiled. The most important thing is to ensure that the liquid comes in contact with the drying agent again and again. Then the liquid is filtered out. If there is formation of any compound due to reaction with water, then the compound needs to be removed by distillation.

2. **Dynamic drying-** The solvent or chemical is passed through a drying tube filled with drying agents. It is important to use such a drying agent that is not susceptible to clumping, as clumping may restrict diffusion and flow rate. Different types of drying agents available for the purpose include silica gel, magnesium perchlorate, molecular sieve, etc. This method improves the rate of drying and it is more effective way of using drying agent.

Techniques of Drying

1. Use of Molecular Sieve

Molecular sieve is an absorbent composed of zeolite which is used for drying gas and liquid as well as other molecules based on the size and shape of the molecular sieve. It works by adsorbing gases or liquids which are smaller than the diameter of pores in the molecular sieve. It excludes (do not adsorb) the gases or liquids which are larger than the pore diameter. Different types of molecular sieves are based on the chemical formula which determines the pore size.

Three different types of molecular sieves generally being used include

- Type 4A: a crystalline sodium aluminosilicate [having pore size of 4 Angstrom (Å)] which can adsorb ethane molecules but not butane besides water).

- Type 5A: a crystalline calcium aluminosilicate (having pore size of 5 Å) which can adsorb molecules greater than 4 Å.
- Type 13X: a crystalline sodium aluminosilicate (having pore size of 10 Å) which can adsorb several branched chain and cyclic compounds.

2. Chemical Drying

In this method a drying agent is used, which is selected based on the product formed in reaction with water that can be easily removed by filtration. An appropriate water-reactive substance (a drying agent) is added to the solvent. Usually, such a drying agent is selected which produce a products on reaction with water that can be easily removed by filtration or distillation. The drying agent commonly used for this purpose include activated magnesium (for alcohols), calcium oxide, calcium hydride, phosphorus pentoxide, and sodium metal (sometimes in combination with benzophenone). Sodium should only be used for clean solvents containing low water content; otherwise there might be a risk of fire or explosion.

3. Azeotropic Distillation

Azeotropic distillation is a method of separating mixture of compounds based on the difference in volatility of a boiling liquid mixture. An azeotrope is a mixture of two or more liquids which produces vapours with the same ratio of constituents as that of the compound of interest; hence they cannot be separated by normal distillation. Therefore, addition of a third compound (entrainer) becomes important. It changes the volatility of one of liquids in azeotrope to a greater extent compared to that of other allowing the separation to occur. Heterogenous azeotrope contains a mixture of constituents that are not completely miscible and present in two liquid phases. Homogeneous azeotrope contains a mixture of constituents that are completely miscible.

4. Salting Out

It is a technique that takes advantage of reduced solubility of a compound in a solvent/solution compared to its solubility in water. This is mostly used for polar solvents that are miscible/highly soluble in water. This technique is often used prior to distilling with a salt, as it cannot remove all the water, but is a convenient way to remove most of it without using any anhydrous compounds. It is one of the ways to break an azeotrope without vacuum distillation. Examples of salting out include the separation of ethanol and water using potassium carbonate, removal of isopropanol from water using sodium chloride, sodium hydroxide, or a mix of these two.

5. Use of Anhydrous Salts/Desiccants

It is the most commonly used method which uses anhydrous form of a salt as desiccant. This method can be used for both polar and non-polar solvents. It involves adding a salt (e.g. anhydrous magnesium sulphate, sodium sulphate, calcium sulphate, calcium chloride) directly to the solvent. Once the salt has been hydrated, it can be taken out by filtration or by decanting, followed by distillation, if necessary.

Naturally occurring gypsum (calcium sulfate) is used as drierite, which is an all-purpose drying agent for efficient and rapid drying of air, gases, refrigerants, organic solids and liquids. It can be used to maintain dry atmosphere in storage spaces, vaults, commercial packagimg as well as other enclosures. It is also used to protect hygroscopic materials and the materials subject to mildew, corrosion, rust, or deterioration caused due to high humidity.

When used for drying of liquids, Drierite instantly absorbs 6.6 weight percent water by chemical action, creating the hemihydrate of calcium sulfate. For drying of gases, drierite has a water capacity of 10 to 14 weight percent. Drierite can be regenerated several times for reuse.

Fig. 2.1: Drierite

Drierite is of neutral pH, constant in volume, chemically inert (except toward water), insoluble in organic liquids and refrigerants, non-wetting, non-toxic, non-disintegrating, non-corrosive, generally recognized as safe (GRAS) by FDA.

Indicating Drierite is impregnated with cobalt chloride, which is blue when dry and changes to pink colour upon absorption of moisture. This makes the

Indicating Drierite valuable to know about its dryness (being maintained) and when the drierite needs to be replaced. Generally it has the same efficiency as regular Drierite but greater capacity due to the desiccating effect of the cobalt chloride impregnation.

6. Freeze Drying

Freeze drying is a process of removing solvent (usually water) from the sample by crystallization at low temperature followed by sublimation. Freeze drying works by freezing the material followed by reducing the pressure in surrounding to make the frozen water (in the sample) to sublimate directly from solid phase to gaseous phase. The sample is frozen below the temperature at which the solid and liquid phases of the material can coexist (normally −50 °C and −80 °C). Then in the primary drying phase, lowering of pressure leads to sublimation of 95% water in the material. Subsequently, in the secondary drying phase the remaining unfrozen water is removed from the sample with residual water content of only 1−4%.

7. Vacuum Drying

Drying is one of the most energy intensive operations because water has high latent heat of vaporization. Though most of the dryers are classified as direct dryers, vacuum drying is another important process wherein drying is performed under lower pressure around the sample. Drying occurs when liquid is removed or vaporized by heating of the solvent/chemical. The bound or unbound liquid is removed by the drying process. The unbound liquid (as a film on upper surface) is easily removed, but bound liquid is present in the micro-structure of the material. Removal of the unbound moisture depends on the external conditions like temperature, flow rate, humidity, exposed surface area and pressure. The removal of bound moisture depends on nature of the sample material and amount of moisture within it. A recent advancement in the process is ***Microwave vacuum drying*** which uses microwave-based heating of the material using electromagnetic field and conversion of electromagnetic energy to thermal energy. It is the fastest drying method and dries the material with less thermal conductivity.

8. Liquid Drying

Liquid dryer is used to remove small amount of water from a solvent. Using a liquid dryer, water is removed by passing the solvent through a freshly regenerated column packed with a desiccant. Each desiccant has specific capacity to absorb water, which requires to be regenerated/replaced time to time. For continuous dryer, dual dryer columns may be used one for solvent streaming and another for regeneration. Thus, drying refers to the use of desiccants to remove excessive amount water from chemical/solvent. Sometimes excess water stops a reaction, even act as a barrier in starting specific reactions like Grignard reaction.

Let's Check Your Understanding

1. What are the two major methods of drying?
2. Which of the following solvents are oxygenated?
    a. Alcohols
    b. Ketones
    c. Glycol ether
    d. All of the above.
3. Which type of sieve is generally used for drying a gas?
4. Which of the following desiccant is available as "Drierite"?
    a. silica gel
    b. alumina
    c. anhydrous calcium sulfate
    d. magnesium sulfate
5. Which of the following is not a property of a drying agent?
    a. Capacity of drying agent
    b. Absorption rate
    c. Surface tension
    d. Intensity of drying
6. What is the characteristic feature of Type 4A molecular sieves?
7. Name two most common anhydrous salts used for drying of both polar and non-polar solvents.

Answers

1. Static drying and dynamic drying
2. d (All of these)
3. Type 5A sieve
4. c (Anhydrous calcium sulfate)
5. c (Surface tension)
6. Crystalline sodium aluminosilicate
7. Magnesium sulphate, sodium sulphate, calcium sulphate.

3

Washing, Drying and Sterilization of Laboratory Glassware

Introduction

A variety of glassware, plasticware, and other apparatus are used for experimental purposes in a laboratory. They need to be washed, cleaned, dried, and sometimes sterilized before they can be used in the next set of experiments. Glassware needs special care of handling during washing, drying, sterilization, etc. because even a minor mistake in handling of glassware may result in its cracking/breakage and it may become unserviceable. All of these may result in a delay in experimentation if spare piece of the glassware is not available or it may affect serviceability of the associated apparatus.

A large number of glassware such as beaker, flask, measuring cylinder, burette, test tube, etc. are regularly used in a laboratory, which requires careful handling during the different stages of experiment as well as after completion of the experiment while preparing them for another set of experimentation.

Glass being a complex silicate, its properties depends on the type of silicate anions in the structure and the cations formed. Thermal properties of glass changes by the addition of boron oxide (B_2O_3); hence, borosilicate glass is extensively used in the laboratories. With this brief background information, let us discuss about the necessity, procedures, and precautions during washing, cleaning, drying and sterilization of laboratory-wares. Figure 3.1 shows various advantages of using borosilicate glass in a laboratory

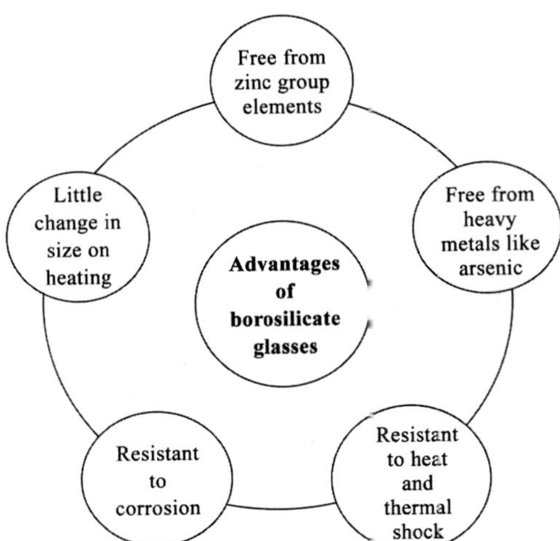

Fig. 3.1: Advantage of using borosilicate glassware

Good laboratory practices require clean glassware, because an experiment may give erroneous result if dirty glassware is used at any step of the experiment. Hence, the glassware must be physically and chemically clean; and in many cases, it must be microbiologically clean or sterile. The commonly used glassware in a laboratory include test tube, measuring cylinder, pipettes, beakers, flasks, microscopic slides, etc.

Washing and Cleaning of Glassware

The cleaning glassware depends upon its type and the cleaning agent to be used for the purpose.

1. Preparatory Cleaning

- Thick solids materials adhering to the glass surface should be removed as soon as possible.
- Greasy material stick at the glass joint should be removed using a solvent like acetone.
- The glassware should be kept in a warm cleaning solution of detergent.
- A brush or cleaning pad may be used to clean any residue or contaminant on the surface.
- The glassware should be rinsed under running tap water, followed by deionized water and then allowed to dry.

- Most of the newly procured glassware is slightly alkaline and should be washed upon receiving. For this purpose, the glassware should be soaked in a 1% HCl or HNO_3 solution, washed with tap water followed by rinsing with deionized water and it should be dried before use.
- In case the glassware doesn't appear to be clean after the initial cleaning, we need to go for more aggressive cleaning protocols as per the requirement.

2. Mild Cleaning

- If the preparatory/preliminary cleaning is not able to remove the dirt, we need to try gentle solvent with soaking.
- While using a gentle aqueous solution, heated soaking and sometimes mechanical agitation like stirring, shaking, or sonication may enhance cleaning efficacy.

3. Cleaning with Organic Solvent

- To remove organic residues from the surface of glassware, it should be rinsed briefly with an organic (acetone or ethanol) solvent.
- Warm tap water or brushing with soapy water should be used to scrub the inner side of curved glassware.
- Soapsuds on the glass surface should be removed using deionized water to avoid water stains. The deionized water is poured on cleaned glass surface, it forms a smooth sheet with uniform spreading. If this sheeting action is not observed, more intensive cleaning might be needed.

4. Aggressive Cleaning

- If the mild aqueous and organic solvent cleaning does not provide the desired level of cleanliness, then aggressive cleaning should be adopted.
- This involves removal of the top layer of silicon oxide of the glassware to scrub out the adhered material/contaminant.
- It can also be done by oxidizing the contaminant by soaking the glassware in 2% hydrofluoric acid or a base-bath (sodium or potassium hydroxide in ethanol or isopropanol).
- The aggressive base-bath cleaning is possible due to the borosilicate nature of laboratory glassware.

5. Cleaning Oxidizing Contaminants

- Sometimes the residue on glassware is insoluble in organic solvent, surfactant solution, or in mild acidic solution.
- In that case, a common ways to clean the glassware is to oxidize the contaminant to make it soluble.
- Oxidizing agent like *Aqua Regia* (a molar ratio of 1 part Nitric acid to 3 parts Hydrochloric acid), Chromic acid (a sulfuric acid-based reagent), Piranha solution (a hydrogen peroxide based agent), and fuming sulfuric acid (containing pyro sulfuric acid) is used for this purpose. However, it must be kept in mind these reagents need to be handled with utmost care.

Safety Precautions During Cleaning

- While cleaning the laboratory glassware, protecting eyes is the first requirement, because even soap solution may cause irritation in eyes.
- Many of the chemicals used in cleaning can easily penetrate the skin, especially when combined with organic solvent. Therefore, chemical splash protective goggles, plastic aprons and non-disposable gloves (specifically designed to handle extended contact with corrosive chemicals) must be used while performing aggressive cleaning.

Checking Cleanliness of Glassware

- Observe the glass surface for the final rinse water drains off.
- The water should move with a sheeting action, leaving a thin film over the whole surface.
- If the film breaks up into droplets or the surface is unevenly wet, it indicate that the glassware is not properly clean.

In this Case, Cleaning can be Performed Again by

- Rinsing it with water, and
- Soak in a detergent solution, followed by washing.

Drying of Glassware

Drying of the glassware's is equally important as washing and sterilization. Use of towel or cloth for drying the glassware may result in appearance of fibres or dusts on the surface, which in turn may affect the sensitivity/accuracy of research work. In order to avoid such conditions, sophisticated methods needs to be adopted for the removal of moisture from the glassware. Some of the most commonly adopted measures are discussed below, in brief.

a. Natural Air Drying

When quick drying is not needed, we can go for natural slow air-drying. Air-drying racks are available nowadays to which we can hang the glassware with their open end facing downward. This helps in preventing the entry of any dust through the opening of glassware/equipment and also helps in quick drying. An important thing to be considered during the drying process is that the rack must be maintained clean free of any contamination.

b. Quick Drying

When immediate drying of glassware is required, we rinse the glassware with acetone and the residual acetone is allowed to vaporize quickly. Since water is miscible with acetone this works well in quickly drying the glassware. However, care must be taken to avoid vaporization of the residual acetone inside a hot oven, because this may result in polymerization of acetone or subsequent ignition.

c. Oven Drying

Sometimes, even the dried glassware may show the presence of a thin layer water film, if we observe the glassware closely, which may result in some experimental error/harmful reaction. Therefore, it may be necessary to remove even a traces of water. For this, we need to keep the glassware in oven at higher temperature (about 110 °C) for several hours or even overnight.

d. Flame Drying

Sometimes, to remove the residual film of water from glassware, flame-drying method is used following the below mentioned steps (Figure 3.2).

(a) Remove the vinyl sleeves from clamps,

(b) Clamp the flask to be dried,

(c) Apply the burner,

(d) Continue heating until the glassware becomes (scorching) hot enough.

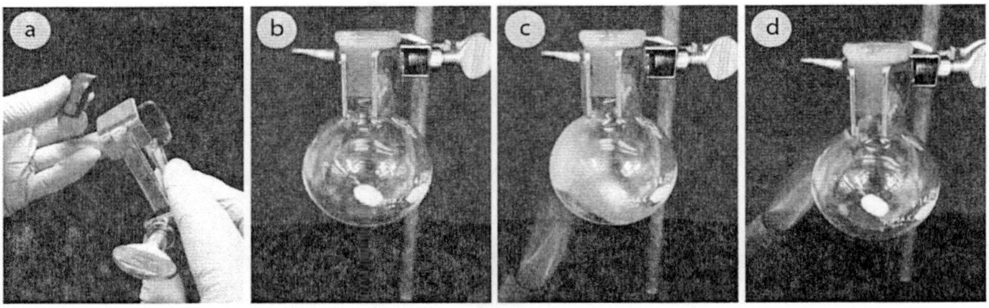

Fig. 3.2: Flame drying of glassware. **(a)** Removal of vinyl sleeves, **(b)** clamping the glassware, (c) flame heating of glassware, **(d)** heating until becomes red-hot.

Sometimes, after flame heating/drying of glassware, we observe the appearance of water vapour along the wall/ surface of the glassware. Such vapour formed after the oven or flame drying can be removed by the following method (Figure 3.3).

(a) A desiccator can be used to remove the trapped water vapour,

(b) A balloon of inert gas (e.g. helium) can be used to remove the trapped water vapour,

(c) A drying tube containing anhydrous calcium chloride ($CaCl_2$) or calcium sulphate ($CaSO_4$) pellet (which readily absorbs water) can be used to remove water vapour trapped inside the glassware.

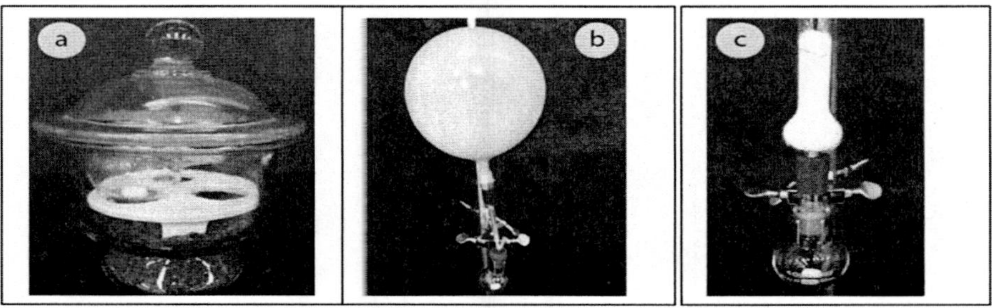

Fig. 3.3: Removal of vapour formed after oven/flame drying. **(a)** Using a desiccator, **(b)** a balloon of inert gas, or **(c)** a drying tube containing drying agents like anhydrous $CaCl_2$ or $CaSO_4$ pellet.

Glasswasher with Built-in Dryer

- Built-in glasswasher and dryer (Figure 3.4) is suitable for high throughput laboratories. This proves to be much efficient, faster, and ensures cleanliness.

Fig. 3.4: Glasswasher with a built-in drying system performs better cleaning and drying process.

Sterilization of Glassware

Sterilization is any process that removes kills or deactivates all microorganisms (including most resistant bacteria and spores) and other biological agents present in/around a specific surface, object or any substance.

Necessity of Sterilization

Microorganisms are capable of causing infections. They are constantly present in the environment and on the glassware in a laboratory. These are responsible for contamination/infection in the cultures. To ensure the glassware in a laboratory to be completely free from contaminations, the best method is its sterilization. The process kills and inactivates the microbes including viruses, bacteria, fungi, and spores.

Factors Affecting Sterilization Efficacy

- Contact time / Time of sterilization
- Temperature of sterilization
- Physico-chemical environment (such as pH)
- Presence of organic materials
- Number of microorganisms
- Material composition of glassware
- Nature of the microbes.

Certain indicators have been developed to monitor the effectiveness of sterilization by measuring various aspects of the process using the mechanical, chemical and biological indicators (Figure 3.5).

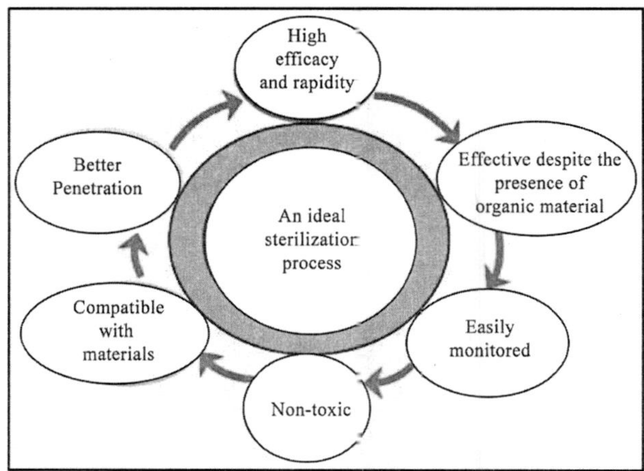

Fig. 3.5: Characteristics of an ideal sterilization process.

Methods of Sterilization

I. ***Physical methods:*** uses physical agents like Sunlight, Heat (Moist and Dry Heat), Filtration, Radiation, Ultrasonic and Sonic Vibrations.

II. ***Chemical methods:*** uses chemical agents like Alcohol, Trichlorobutanol, Aldehydes, Formaldehyde, Dyes, Halogens, Phenols, Surface Active Agents, Metallic Salts, Gases, etc.

Glassware is usually sterilized by

a. Dry heating – Hot air oven

Hot air ovens are the electrical devices used for sterilization by dry heating the articles to be sterilized (Figure 3.6). These are usually operated at 300 °C (572 Degree Fahrenheit). Usually, there is a thermostat to control the oven temperature. Materials being sterilized must be properly arranged to allow free circulation of hot air.

Fig. 3.6: Hot air oven used for heat sterilization of glassware

Recommended temperature and holding time

Temperature (Degree Celsius)	Heating/holding time (minute)
160	45
170	18
180	7.5
190	1.5

b. Moist heating –*Autoclave / Steam Sterilizer*

Autoclave is a pressurized device designed to kill/destroy microorganisms and spores using high temperature at high pressure through steam sterilization (Figure 3.7). Water boils at 100 °C under normal atmospheric pressure. But in an autoclave, water is heated in a sealed vessel which leads to increased pressure hence the temperature inside the vessel increases. The temperature at which water boils also increases. Steam (at higher temperature) having a better penetration power kills most of the microbes. Generally, autoclaving temperature is maintained at 121 °C for 18 minutes at 15 pounds of pressure per square inch.

Thermal death time (TDT) is the minimum time required to kill a suspension of organisms at a predetermined temperature in a specified environment. It is inversely proportional to temperature. TDT increases in presence of organic substance, proteins, nucleic acids starch, gelatin, sugar, fats and oils.

Advantages of Sterilization

- Destruction of resistant bacterial spores with relatively brief exposure.
- No toxic residue/materials after sterilization.

- No side effects.
- Dry-heat sterilization is an easy and economical
- method.

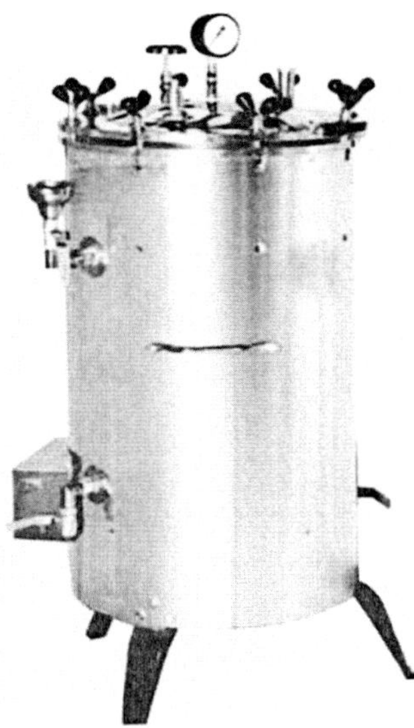

Fig. 3.7: An autoclave used for steam sterilization.

Recommended Temperature and Duration for Steam Sterilization

Temperature (°C)	Duration (minutes)
121	15
126	10
134	3

Thus, even a most carefully executed piece of experiment may give an erroneous result if dirty glassware is used. Therefore, proper washing/cleaning of glassware is one of the most important steps in any laboratory analysis. The procedures discussed above (such as proper washing, drying and sterilization of glassware) are the prerequisites for success in laboratory experiments.

Let's Check Your Understanding

1. Choose all the common physical agents used for sterilization process in a research laboratory, hospital, and biotechnology industry. Select all that apply:

   a. Dry heat
   b. Radiation
   c. Steam heat
   d. Phenols

2. Which of the following is a recommended temperature and time periods for moist heat sterilization using an autoclave?

   a. 180 °C for 5 minutes
   b. 121 °C for 15 minutes
   c. 126 °C for 3 minutes
   d. 160 °C for 45 minutes

3. Name the sterilization agent that is most frequently used for a heat-labile liquid substance or antibiotics

   a. Dry heat
   b. Radiation
   c. Filtration
   d. Formaldehyde

4. Which of the following methods is safe and effective in the elimination of COVID-19? Select all that apply:

   a. Use of 70 % Ethyl alcohol
   b. Use of antibacterial ointments
   c. Hand wash with soap and water
   d. Intake of antibiotics as soon as a symptom appears

5. All of the following chemical disinfectants used in laboratories and healthcare industries have been found to be effective against many bacteria, fungi, and viruses, except

   a. Alcohols
   b. Ethylene oxide
   c. Formaldehyde
   d. Steam heat

6. What is the criterion of selecting a detergent to clean different laboratory glassware?

   a. Economical
   b. Renowned brand
   c. Easily available in the market
   d. Compatibility with supplied water and experiment.

7. Which chemical was first introduced by Joseph Lister for its use as antiseptics in hospital?

   a. Ethyl alcohol
   b. Carbolic acid
   c. Hydrogen peroxide
   d. Chlorohexidine

8. Which temperature and time period commonly used for hot-air oven sterilization of glassware in a laboratory?

   a. 180 °C for 30 minutes.
   b. 63 °C for 30 minutes.
   c. 121 °C for 15 minutes.
   d. 160 °C 45 minutes.

9. Name the chemical which is an active ingredient of 'bleach', a household decontamination product used to kill bacteria, fungi, and viruses?

   a. Sodium chloride
   b. Ethylene oxide
   c. Sodium hypochlorite
   d. Ethyl alcohol

10. Which one should be used for final rinsing of glassware?

   a. Salt water
   b. Diluted acid
   c. Deionized water
   d. Sodium hydroxide

Answers

1. a, b, and c
2. b
3. c
4. a and c
5. d
6. d
7. b
8. d
9. c
10. c

4

Use and Handling of Specific Glassware in Laboratory

A variety of glassware like burettes, pipettes, measuring cylinders, flasks, separating funnel etc. are used for experimental purposes in a laboratory. All of these need to be used and handled correctly, otherwise the chances of experimental error and damage to the glassware increase significantly. Being made of glass, they need special care while washing and using in experiments because even a minor mistake in handling of glassware may result in breakage of the apparatus. With this precautionary note, let us discuss about some of the specific glassware used in the laboratories.

Burette

Burette is a long graduated glass tube with a stopcock at one end for delivering a known/desired volume of a liquid, especially in titration experiments. It contains a stopcock at the lower end and a tapered capillary tube as the stopcock's outlet. The flow rate of liquid from the tub stopcock's outlet can be controlled with the help of stopcock valve. Burette is used for measuring the volume of a solution/chemical added in a quantitative (titrimetric) estimation experiment.

Procedure of using a Burette

The burette reading must be noted before and after delivering the chemical/ solution from the burette. The difference in these two readings is the volume of the chemical/solution released out of the burette. The solution should be released slowly/drop-by-drop to avoid any experimental error/mistake. If the solution/chemical is allowed to pour out too fast, the wall of the burette will not drain properly and some liquid may remain sticking to the surface of the walls. Moreover, excessive amount of chemical may be released/dropped during the experiment, particularly during the titration experiment wherein we need to know the exact amount of acid/base needed to get neutralization point. Thus, any minor mistake may lead to get a faulty reading. The measuring capacity of a burette usually used in a laboratory is 50 mL.

Types of Burette

Generally, two different types of burettes are used in the laboratories: (i) Volumetric burette- it is used to deliver a measured/desired volume of liquid, (ii) Digital burette or Piston burette is similar to a syringe with precise bore and plunger (Figure 4.1). Piston burettes may be manually operated (hand cranking or spinning) or it may be digitally motorized (electronically through batteries). Digital burette may possess a LCD screen, light protection, recirculation system and/or computer interfaces for data logging. The digital burette uses a high-precision syringe to precisely deliver a measured/desired volume of sample. The volume of sample delivered by a digital burette is displayed on a digital display (LCD screen), which eliminates the reading error due to meniscus formation, etc. The plunger/piston moves incrementally by hand turning a ratcheted wheel or by a step motor so that to precisely deliver the desired/ measured aliquot. The barrel and plunger of a digital burette are typically made of glass, however plastic barrel and plunger are used for making the burette to be used for handling of alkali solutions, which may corrode glass.

Applications of Burette

- Beneficial for use in titration,
- Burette is used to dispense small volume of liquid (aliquots) or sometimes gas with higher accuracy.
- Digital burette ensures accuracy required in research and industrial titrations, environmental fieldwork, and wastewater treatment applications.

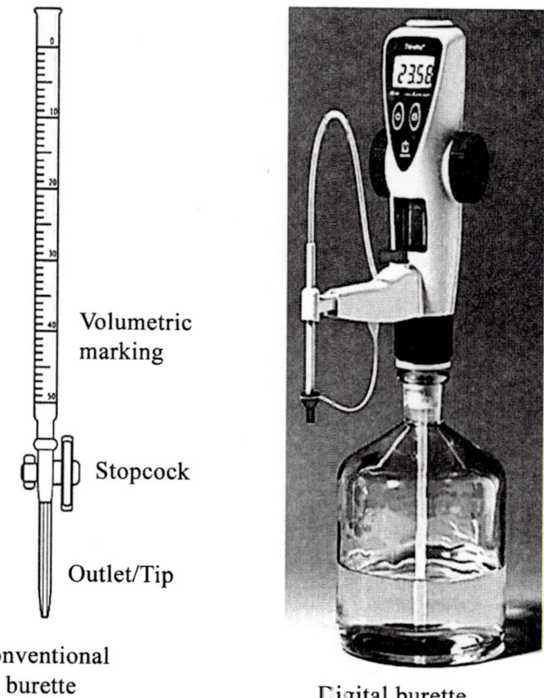

Fig. 4.1: Different types of burette

Applications of Burette

- Used in titration experiments.
- Used to dispense small volumes (aliquots) of liquid, or sometimes gas, with accuracy.
- Digital burette ensures accuracy for industrial titrations, environmental fieldwork, general chemistry and wastewater treatment applications.

Precautions in Handling a Burette

- The initial and final burette reading must be noted before and after delivering the liquid chemical.
- The difference between these two readings (Final – Initial reading) is the volume of the liquid chemical delivered.
- The chemical/solution should be delivered drop-by-drop.
- Before filling the burette with liquid chemical/solution to be used, the burette should be rinsed with the solution.

- For rinsing the burette, a few mL of solution is taken into the burette and inner surface of the burette bust be wetted with the solution by rotating the burette.
- After properly wetting the inner surface, the solution must be drained out through nozzle of the burette.
- After rinsing, the solution should be filled in the burette with the help of a glass funnel above the zero mark. Stopcock should then be opened wider so that the solution is allowed to run through the nozzle until no air bubble is trapped in the nozzle.
- After filling/refilling the solution/chemical in the burette, the solution to be titrated must be placed below the burette nozzle and the solution/ chemical in burette must be released slowly (drop-by-drop) with the help of the stopcock which mixing the contents.
- Dropping of the solution/chemical should continue until the end point reaches, which needs to be detected based on certain visible parameter(s). For example, in a neutralization reaction, end point can be detected by change in colour of the solution [red colour of an acidic solution becomes colourless when it gets neutralized (becomes colourless)].
- In order to measure the volume of solution released out of the burette, the reading at the wall of burette parallel to the lower surface of the meniscus must be noted down as the final reading.
- In case of using a light coloured solution taken in burette, anti-parallax card (or any other dark colour paper sheet) may be placed behind the burette filled with solution (at the point of meniscus formation), which may help observing the lower surface of the meniscus formed by the solution in the burette.
- In case of very dark solution like potassium permanganate, measurement should be taken at the upper layer of meniscus formed.
- It must be ensured that no liquid solution/chemical has leaked out.

Operating the Burette

Proper burette technique is an important laboratory skill for accuracy in laboratory findings. A right-handed person should operate the burette with the left hand, and a left-handed person should operate the burette with the right. This enables more coordinated handling in swirling the reaction flask, if needed.

- Before delivering any solution from the burette, initial burette reading must be recorded.
- Open the stopcock by twisting it 90 degrees to vertical position and allow the solution to drain drop by drop (slowly).
- As the end (neutralization) point approaches, the flow rate should be further slowed (one drop at a time) by turning the stopcock back toward the closing position.
- When the end (neutralization) point reaches (the desired volume has been delivered), close the stopcock and wait for a couple of seconds, then record the final burette reading.
- Calculate the volume delivered by subtracting the initial reading from the final reading:

Volume delivered = Initial reading − Final reading.

While delivering the solution, one must not allow the solution to drain below the bottom of the calibration range on the burette. When this is about to occur, close the stopcock, take the reading, refill the burette, take the initial reading, and continue delivering the solution. The total volume delivered would be the sum of the volumes delivered at the first time and the volume delivered in the second step.

Reading a Burette

While noting down the burette reading, one must ensure that the solution level (meniscus formed by the aqueous solution) is clearly visible. If needed, anti-parallax card can be used as shown in the Figure 4.2. Moreover, while recording the reading on burette the level of eye must be in the horizontal position (Figure 4.3).

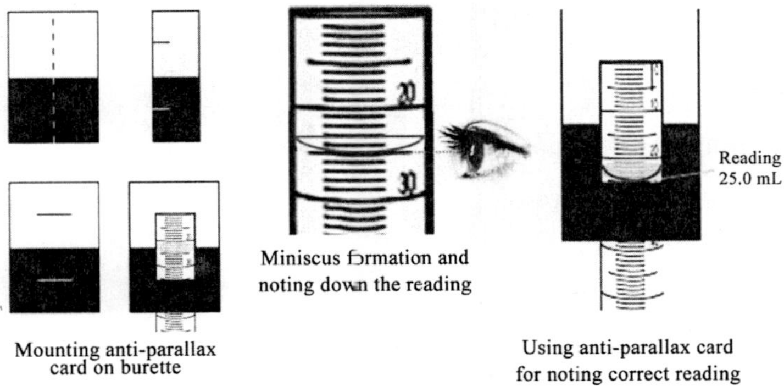

Fig. 4.2: Method of taking/noting reading on a burette.

Sources of Error in Burette use

1. Air trapped in the stopcock or burette tip.
2. Parallax error: This type of error occurs when the scale of the burette is not viewed from a perpendicular position. Looking from lower position of eye (down on the meniscus) causes it to appear higher than where it really is. Looking up at the meniscus causes it to appear lower than it really is.
3. Delivering the liquid too rapidly, so that drops form on the side of the burette.

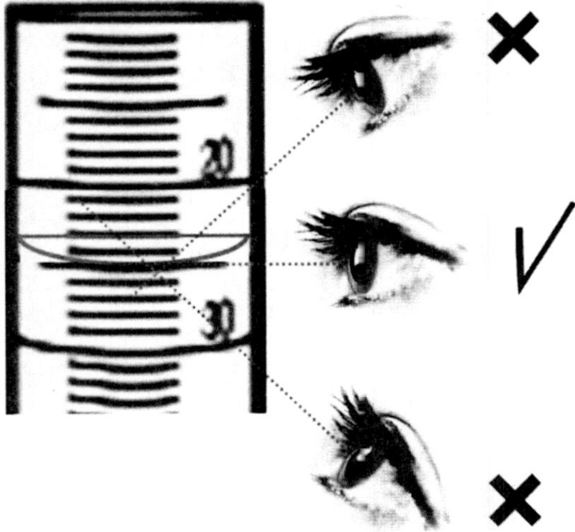

Fig. 4.3: Parallax error in reading a burette

Pipettes

Pipette is used in a laboratory to measure or transfer small quantity of liquid (milliliter) for chemical, biochemical, and medical experiments (Figure 4.4).

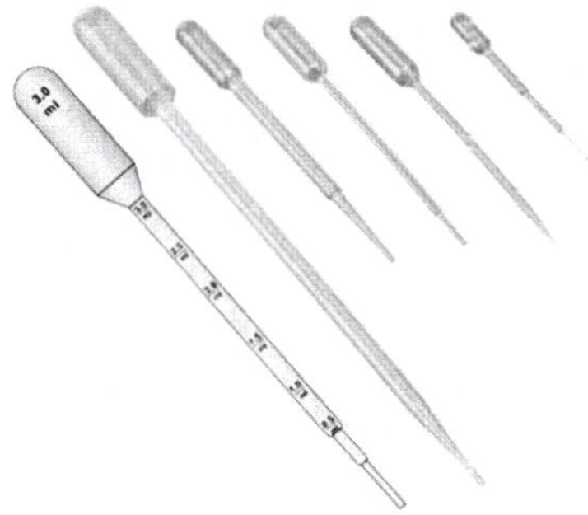

Fig. 4.4: Pasture pipettes of varying size.

Types of Pipettes

Depending on the type and usage, pipettes can be of the following types.

1. ***Pasteur pipette:*** Pasteur pipettes commonly used now-a-days is generally made of plastic and used to transfer small amount of liquids. It may or may not be graduated and may be supplied as sterile (packed inside a polypack) or non-sterile pipettes.

2. ***Serological pipettes*** Serological pipette is used in cell and tissue culture applications when 1.0 or >1.0 mL volume is to be pipetted (Figure 4.5).

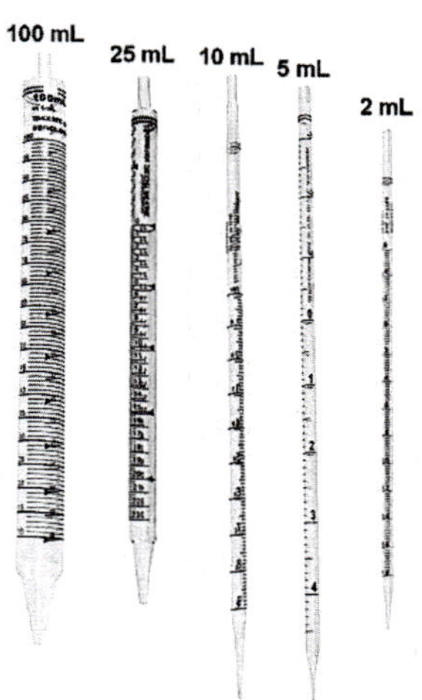

Fig. 4.5: Serological pipettes.

Serological pipettes are made of glass or plastic. Plastic pipette is disposable and sterile plastic pipette is used in tissue culture or when sterility is to be maintained. Glass made serologibal pipettes are reusable and they can be sterilized as and when required using a steel containing (Figure 4.6).

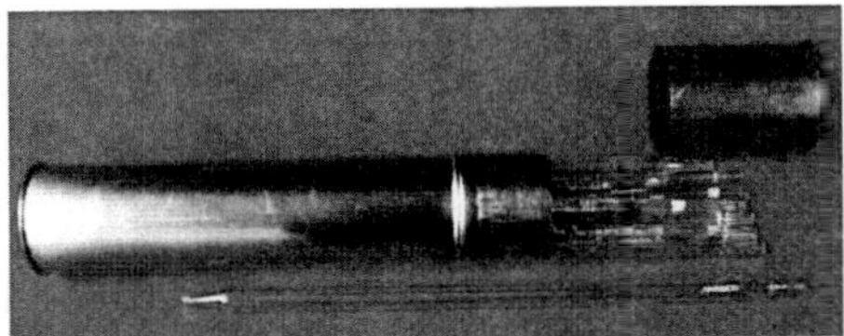

Fig. 4.6: Steel case used to sterilise serological pipettes.

Care to be Taken While using Pipette

- Avoid mouth pipetting.
- Before using a pipette, check the capacity of the pipette, which should be of the desired size/volume.
- Make yourself familiar with the marking/divisions marked on the pipette before dispensing the liquid from it.
- Use a pipette filler bulb to draw the liquid up into the pipette.
- Rinse the pipette with the solution which is to be measured using it.
- For rinsing, take a few milliliters of the solution into the pipette and wet the inner surface of the pipette with the solution by rotating it.
- After rinsing, drain out the solution completely from pipette nozzle (for this a pipette bulb can be used, Figure 4.7).

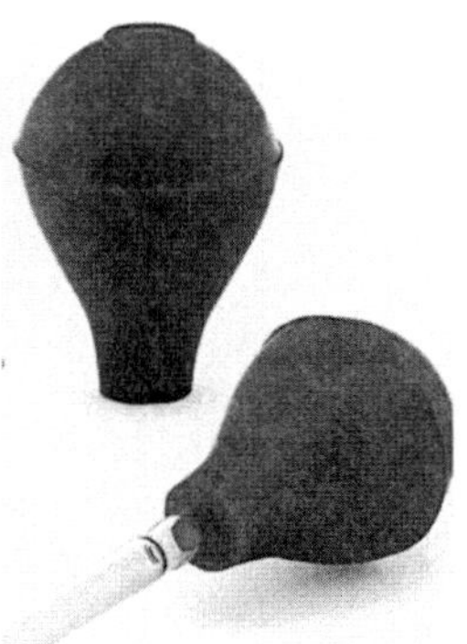

Fig. 4.7: Pipette bulb.

- Pipette is now ready for measuring/transferring the solution.
- At the time of handling a pipette, hands should be dry so that release of air/pressure is regulated easily.
- Nozzle of the pipette being used should not be broken. Avoid the use of broken pipettes.
- Do not try to blow out the remaining liquid (if any) present in the nozzle of the pipette.

Measuring Cylinder

Like weighing, measuring the volume of liquid is an essential operation and a frequently encountered process in a laboratory. A measuring cylinder, also known as graduated cylinder, is standard laboratory equipment used to make accurate volumetric measurements (Figure 4.8).

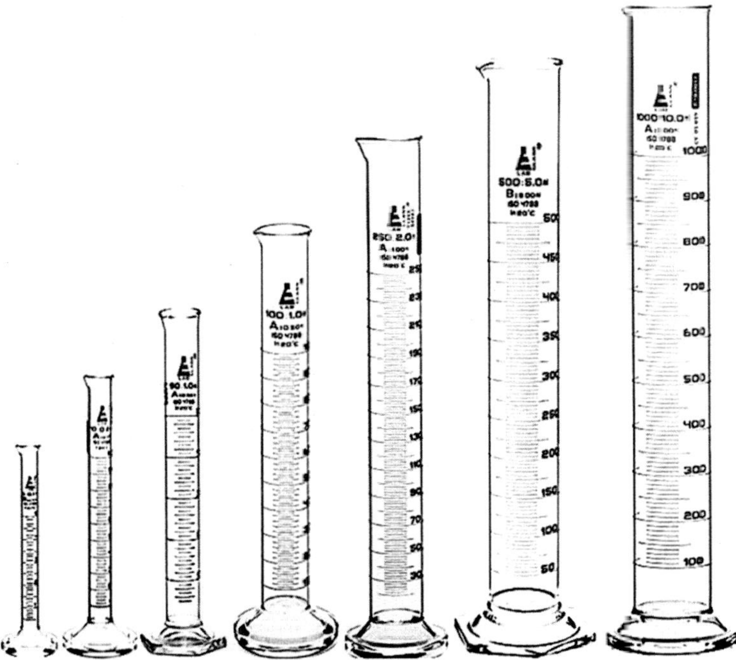

Fig. 4.8: Measuring cylinders of variable sizes (10, 25, 50, 100, 250, 500, 1000 mL).

A measuring cylinder is a cylindrical glass or plastic tube that is sealed at one end. On the outer wall, graduations are engraved after calibration at a specific temperature, (usually at 25 °C). They come in different size or volume capacity, typically ranging from 10 mL to 2000 mL.

Measuring cylinders serve to measure different amounts of liquid. Therefore, the materials used to make the measuring cylinder are also different. High-grade borosilicate glass is often used, which is more durable and easier to read the graduation marks, compared to that made of plastic. However, borosilicate glass cylinder is also having some disadvantages like being easier to be broken and a bit heavier to handle.

Plastic measuring cylinder is made of clear polymethylpentene (PMP). They are shatterproof and they can be steam sterilized (autoclaved) at 121 °C. However, the PMP cylinder is more brittle than polypropylene cylinder.

The size of measuring cylinder determines the scale/divisions made on a graduated cylinder. A 50 ml graduated cylinder is divided into 1.0 ml increments. However, the scale of a 10 mL graduated cylinder is divided into 0.1 mL increments. Similarly, the scale of a 500 mL graduated cylinder is divided into 5 mL increments.

Small sized measuring cylinders (.g. 10, 25 mL) forms meniscus. While making the meniscus, the curve in the upper surface of the liquid is close to the cylinder wall. The meniscus curvature depends on the surface tension of the liquid and it is inversely proportional to the diameter of the tube/cylinder in which it is formed. The meniscus can be either concave or convex, which depends on the liquid and the material used for making the cylinder. The meniscus formed by most of the liquids is concave-up, with the lowest point in the centre is used to determine the meniscus reading. The line of sight must be at the same level as the bottom of the meniscus to correctly read the volume.

The meniscus formed by non-wetting liquid (like mercury, Hg) is convex, with the highest point in the centre. In the case of a convex meniscus, the highest point is used for correct reading. While reading any meniscus, it is crucial to ensure that it is in an equilibrium position.

Flask

A laboratory flask is a lab vessel with wide body and narrow, tubular neck. They are generally used for mixing, measuring, and heating chemicals, samples, and solutions. According to the requirements, different types of flasks are available which can be selected and used for the purpose. Laboratory flasks can be made of glass or plastic and they are of different sizes (of a specified volume). Generally, a stopper/cap is provided to seal the opening of the flask.

Uses

- Flasks can be used for making a solution, holding, or collecting it.
- Volumetric measurement of chemicals, samples, solutions, etc.
- For chemical reaction or other processes such as mixing, heating, cooling, dissolving, precipitation, boiling etc.

Types of flasks

A. Conical (Erlenmeyer) flask: It was named after the German chemist Emil Erlenmeyer, who first prepared and used the flask in 1860. It was designed in such a way that mixing/swirling of its contents can be easier, without any spill out. This also makes it useful for boiling liquid in it. Moreover, its longer neck supports holding filter funnel. An Erlenmeyer flask is a cone-shaped container with a neck to easily hold the flask or attach to a clamp (Figure 4.9). The conical shape makes it stable. Most of the flasks are made of borosilicate glass so that they can be heated over a flame or autoclaved to sterilize them. The flask can be of varying size/capacity (25, 50, 100, 250, 500, 1000 mL, etc. The flask can be sealed with a cork or stopper or plastic/paraffin film or a watch glass on top of the neck. Flask is generally used for laboratory experiments to mix different chemicals or to titrate a solution. They are also used in recrystallization experiments as hot vapour condenses on the upper section of the Erlenmeyer flask, reducing the solvent loss.

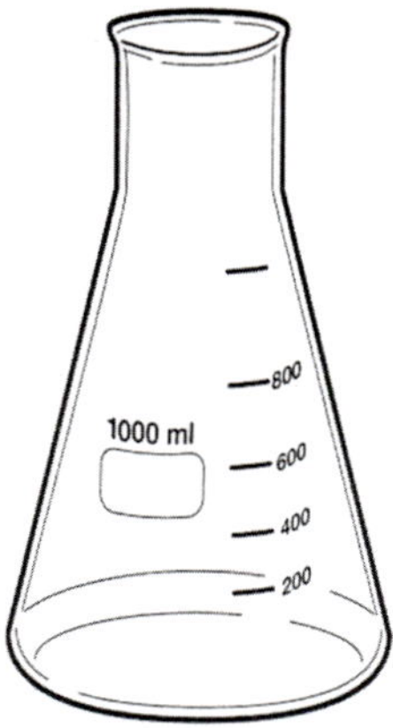

Fig. 4.9: Conical (Erlenmeyer) flask.

B. Volumetric flasks: Volumetric flask is used primarily for the preparation of standard solutions. A volumetric flask is also known as graduated flask. It is a pear-shaped vessel with a very long narrow neck having flat bottom, designed to hold a specific volume at a particular temperature (Figure 4.10). The vessel is indicated with the nominal volume, manufacturer, tolerance, precision class, temperature and manufacturing standard. Volumetric flasks are manufactured in a range of sizes from 25 to 5,000 mL of liquid. A thin circle etched on the neck of the flask indicates the volume of the liquid up to which it holds the volume at a definite temperature (25 °C).

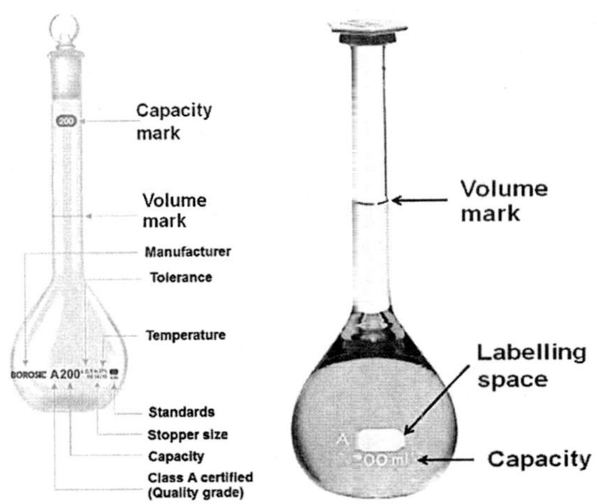

Fig. 4.10: Volumetric flasks.

C. Round-bottom flask: It looks similar to a volumetric flask, but there are differences like thicker neck and no marking (circle etched on the neck) for volume make up (Figure 4.11). Its round bottom helps uniform spread of heat when heated. Round-bottom flask may have a ground glass joint on their neck, to allow them to connection with other apparatus. It is also known as Florence flask, named after Florence in Italy. It possesses a slight lip or flange around the tip of the neck. Such a flask also comes with a flat bottom, so that it can be free-standing (Figure 4.12).

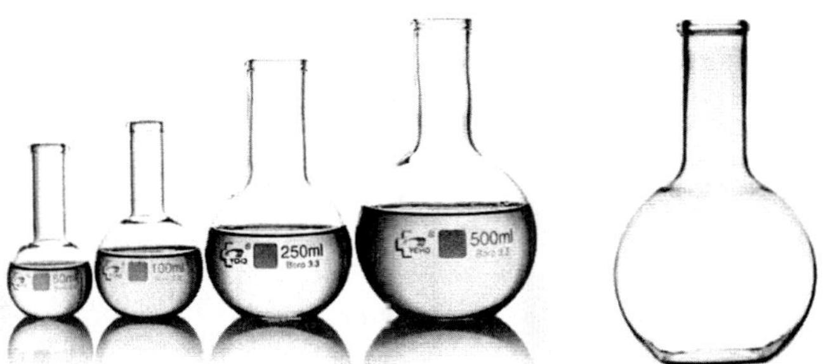

Fig. 4.11: Round-bottom flasks.

Fig. 4.12: Flat-bottom flask.

D. Distillation flask: Distillation flask is used to distil a mixture or to receive the product of distillation. They might be spherical or pear-shaped, sometimes also known as Kjeldahl Flasks, due to their use in Kjeldahl apparatus (Figure 4.13). However, a Kjeldahl flask has a longer neck, which is used to determine the nitrogen content in a sample.

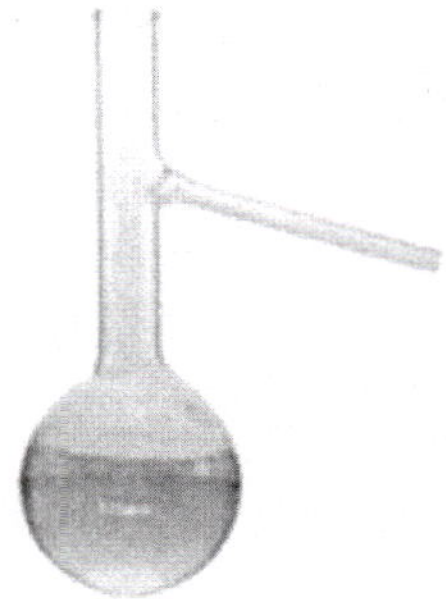

E. Pear-shaped flask: is smaller in size, used for small-scale distillation. Its shape allows recovery of more material than other (e.g. round-bottom) flasks. Other odd-looking flasks have been used for distillation primarily before the advent of condenser.

Fig. 4.13: Distillation flask.

F. Schlenk flask and Straus flask: look fairly similar but Schlenk flask is commonly used in air-sensitive chemistry, as its side arm allows an inert gas (e.g. nitrogen) to be pumped into the flask. On the other hand, Straus flask is used to store dried solvents. The main neck is filled in halfway up, and connected to a plugged smaller neck. The main neck can be connected to other apparatus, and allows the solvent to be extracted when the plug is slightly withdrawn or removed.

G. Claisen flask: was designed for vacuum distillation by chemist Ludwig Claisen. Distillation under vacuum produces problematic amounts of bubbles when solutions are boiled. Claisen's flask includes a capillary tube that inserts small bubbles into the liquid, easing the ferocity of boiling, while the branched portion of the flask hosts a thermometer. It is less commonly used now.

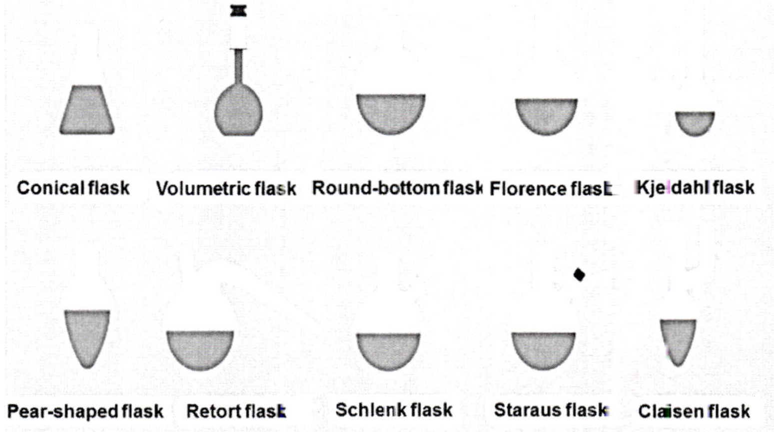

Fig. 4.14: Different types of flasks.

h. Buchner flask: is also known as Sidearm flask or Suction flasks. It is a flat-bottom flask composed of thick, resistant glass. It has a cone form (similar to an Erlenmeyer flask) with a side arm on the neck (Figure 4.15). The flask is used for vacuum filtration, to work with vacuum aspirators or vacuum pumps.

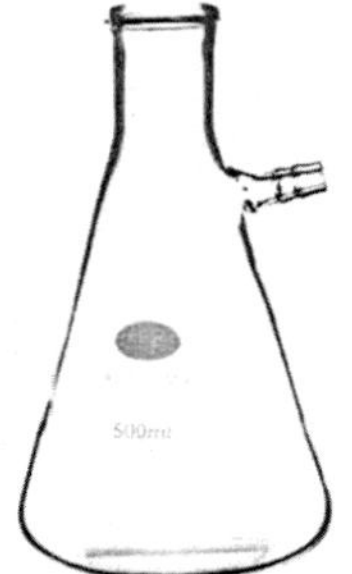

Fig. 4.15: Buchner flask.

Separating Funnel

A separating funnel (also known as separation funnel or separatory funnel) is laboratory glassware used in liquid-liquid extractions to separate (partition) the components of a mixture of two immiscible solvent phases having different densities. When two immiscible liquids are placed/positioned in a separating funnel, two layers are separated. A denser solvent (like halogenated solvents) forms the bottom layer. A separating funnel has the shape of a cone surmounted by a hemisphere. Separating funnel has a stopcock at the bottom that can be opened or closed to drain out the lower layer, as well as a tap at the top which can be opened to release excess vapour pressure. The funnel is typically made of borosilicate glass, which is stronger than regular glass and resistant to thermal shock (Figure 4.16).

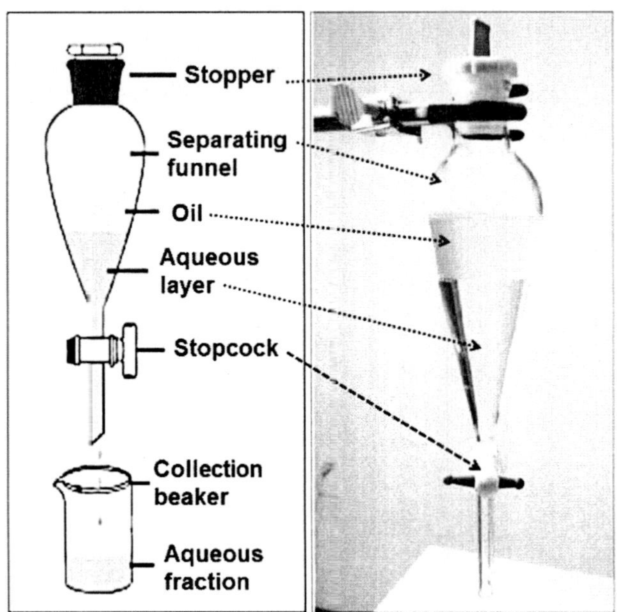

Fig. 4.16: Parts of separating funnel

Parts of a Separating Funnel

1. **Stopper**: It fits tightly, so that the solution does not leak out when the separating funnel is inverted.
2. **Stopcock**: A small amount of grease can be applied to the glass plugs to facilitate sealing. The grease must be used sparingly, because excessive use if grease may clog up the hole in the plug. However, a Teflon plugs possess threads that allow easier opening and release of liquid, it need not be lubricated (Figure 4.17).

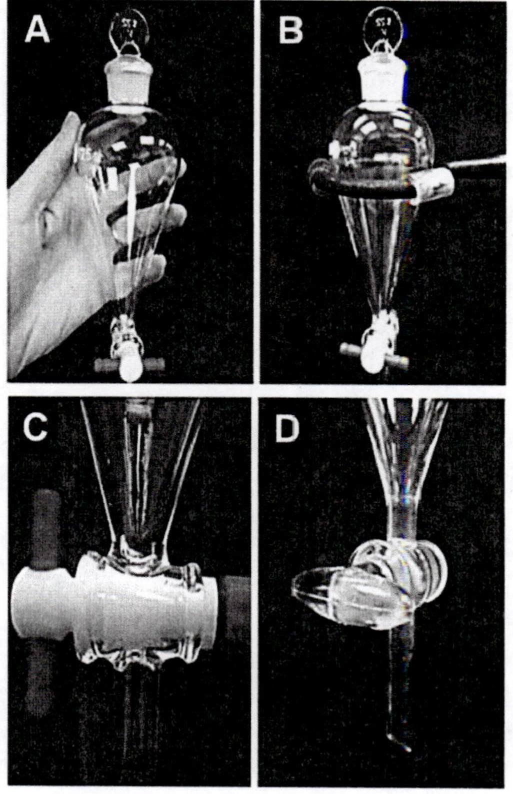

Fig. 4.17: (A) Separating funnel, **(B)** Separating funnel in cushioned ring clamp, **(C)** Opened orientation of teflon stopcock components, **(D)** Glass stopcock at closed

Handling of Separating Funnel

a. Before using

Suspend the separating funnel in an iron ring. While the stopcock is closed, pour ~20 mL of water into the separating funnel (using a short stem funnel).

If the solvent does not leak out at the stopcock, place the stopper on top and invert the funnel. If the liquid leaks, place the funnel back in the ring stand and remove the stopper. Open the stopcock and drain the solvent. If the grease clog is formed and restrict the draining of solvent, use a wooden applicator or piece of wire to open it. There is no need to dry the separating funnel.

b. Performing an Extraction

- Place the separating funnel in an iron ring.
- Remove the stopper and make sure that the stopcock is closed.
- Add the mixed solution to be separated/extracted using a short-term funnel.
- Do not fill the funnel more than half.
- Add the solution mixture and place the stopper to close.
- Keep the separating funnel in the stand and allow the layers to be separated.
- Remove the stopper, drain the bottom layer first into a clean container. We need to know which layer upper or lower) contains the product of interest.

Usages of Separating Funnel

Separating funnel is used to separate compounds via extraction. This approach can be used with two immiscible solvents (i.e. the solvents that do not mix), one of which is almost invariably water. Ethyl acetate, hexane, chloroform, dichloromethane, and diethyl ether are common organic solvents that provide a clear separation between the two liquids. The aqueous phase and the organic phase are the two layers that are often referred here. It's critical to keep track of these phases because their positions are affected by the solvent. The organic phase floats on the top of separating funnel as it is lighter than water (density 1.0), whereas the solvent denser than water (density >1.0) will form the bottom layer.

Let's Check Your Understanding

1. Read the statements and choose the right option:

Statement 1: Polymethylpentene measuring cylinders have higher chemical and heat resistance than Polypropylene measuring cylinders but tend to be more brittle.

Statement 2: Polypropylene measuring cylinders have higher chemical and heat resistance than Borosilicate measuring cylinders but tend to be more brittle.

   a. Both statement 1 and 2 are true

   b. Both statement 1 and 2 are false

   c. Statement 1 is true and statement 2 is false

   d. Statement 1 if false and statement 2 is true.

2. The principle of separating funnel technique is:

   a. Liquids separate out in layers depending on their solubility.

   b. Liquids separate out in layers depending on their density.

   c. Liquids separate out in layers depending on their melting points.

   d. Liquids separate out in layers depending on their boiling points.

3. Volumetric flasks are used as distillation flasks in Kjeldahl apparatus. **True / False**

4. While taking reading on a burette, what is the best practice in general?

   a. Read the bottom of the meniscus

   b. Read the top of the meniscus

   c. Estimate the last decimal place

   d. None of these.

5. Which is the method used for cleaning a burette?

   a. Rinse with distilled water   b. Rinse with solution

   c. Use wire to clean out the nozzle   d. All of the above

6. Which burette, used in laboratory, has the highest resolution?

   a. 50mL   b. 30mL

   c. 20mL   d. 40mL

7. You must immerse the barrel of the micropipette in the fluid, when taking the liquid.

True/False

8. What is the alternative term used for Graduated pipettes?

   a. Serological pipettes b. Mohr pipettes

   c. Both a and b d. None of the above.

9. Pipette preferred for volatile or viscous substances or measuring a small volume with highest precision is

   a. Volumetric pipette b. Serological pipette

   c. Positive displacement pipette d. Pasteur pipette.

10. Identify the position of the stopcock of a Separating funnel

    a. Open position b. Close position

    c. Continuous flow position d. None of the above.

Answers

1. c
2. b
3. False
4. a
5. d
6. a
7. False
8. c
9. c
10. b

5

Use and Handling of Laboratory Equipment

Different basic as well as advanced laboratory equipment need to be used in a laboratory while conducting experiments. Many of them like pH meter, laminar air-flow hood, viscometer, thermometer, magnetic stirrer, microwave-ovens, incubators, sand-bath, water-bath, oil-bath, etc. are the basic and simple equipment which require to be used routinely while conducting experiments in a laboratory. However, basic knowledge of their operation, handling procedure, when to be used, how to be used, and more importantly what not to be done with it, etc. are essential for keeping them in serviceable condition as well as for the accuracy/precision in their operation. In this chapter, we will discuss about some of the routinely used laboratory equipment along, while some other sophisticated equipment will be discussed in other chapters.

pH Meter

A pH meter measures exactly the H^+ concentration of a solution; thus indicates acidity or alkalinity of a solution with greater precision. Actually, it measures the difference in electrical potential between a pH electrode and a reference electrode, and display the results converted into the corresponding pH value. Electrode is the key part (rod-like structure, usually made of glass) of a pH meter, having a bulb at the bottom containing a sensor. The pH meters range from simple and inexpensive pen-like devices (portable/pocket pH meter) to complex and expensive laboratory instruments with computer interface (Figure 5.1).

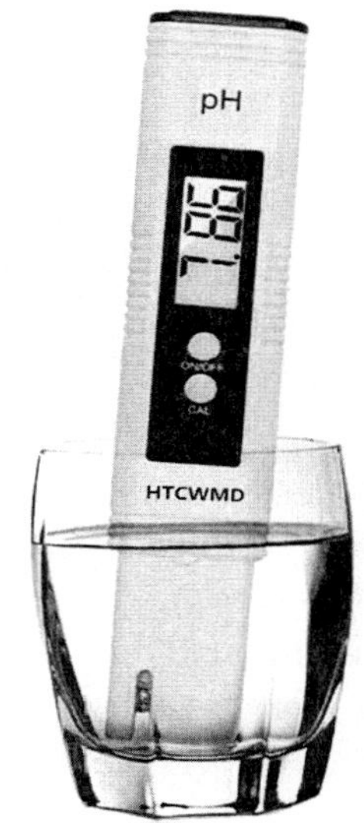

Fig. 5.1: Portable/pocket pH paper.

Frequent calibration of a pH meter using the solutions of known pH (standard solution of pH

4.01, 7.00, 10.01, Figure 5.2), generally before each use, ensures the accuracy of the instrument. For very precise work the pH meter should be calibrated before each measurement. For normal use calibration should be performed at the beginning of each day. This is necessary because the glass electrode does not give a reproducible measurement over a longer period of time.

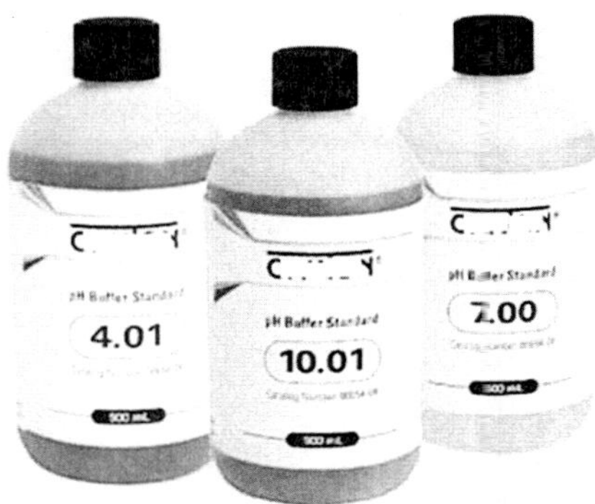

Fig. 5.2: Standard pH solutions

Calibration should be performed with at least two standard buffer solutions that span the range of pH values to be measured. For general purposes buffers of pH 4.00 and pH 10.00 are acceptable. Standard pH buffers can be purchased from the market (commercially available) or else they can also be prepared by dissolving pH tablets (commercially available, Figure 5.3) in a specified volume of distilled water (as per the instructions provided by the manufacturer/ supplier of the pH tablets. However, the standard pH buffers should be stored at 4 °C in a refrigerator after use, and they can be used several times, before they get expired (as mentioned by the manufacturer) or they get turbid/ contaminated or up to 3–6 months. For calibration of the pH meter (Figure 5.4), the instructions/steps given by the manufacturer of the instrument must be followed, as this may vary from instrument to instrument. For calibration of pH meter, generally a button for calibration is there to set the pH meter reading equal to the value of the first standard buffer. Similarly, another button is there to adjust the meter reading to the value of the second buffer. Another button helps adjusting the temperature. For more precision in measurements, three buffer solution calibration is preferred. As pH 7 is a "zero point" calibration (zeroing or taring a scale), calibrating the meter first at pH 7.0, followed by calibrating for the pH value (4 or 10) closest

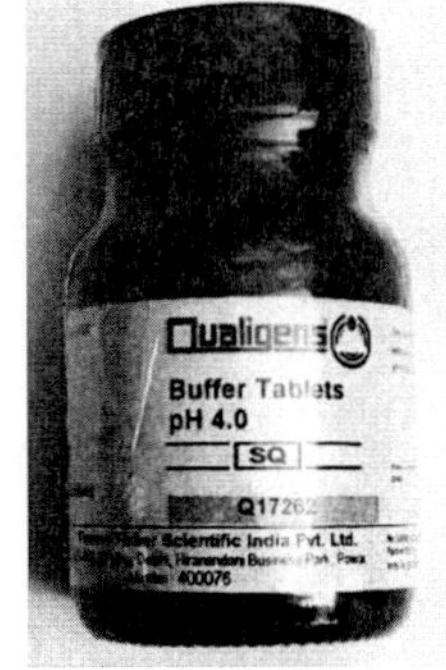

Fig. 5.3: pH tablets.

to that of the test solution, and finally calibration for the third point (10 or 4) provides better linear accuracy to the pH meter. After each step of calibration, the probe must be rinsed with distilled water to remove any trace of the buffer solution adhering to the electrode, followed by blot drying with a clean wipe to absorb the remaining water.

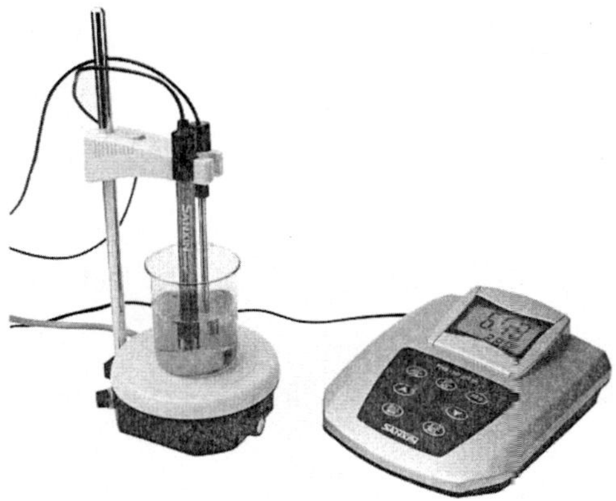

Fig. 5.4: pH meter

Steps in Calibration of pH Meter

1. Before you begin to calibrate and/or use pH meter, you will first need to turn-on the pH meter to warm it up for an adequate (30 minutes) period. However, you must check the operating manual provided with the pH meter.

2. Take out the electrode from the storage solution and rinse it with distilled water keeping an empty beaker below to collect waste. Once rinsed, blot dry the electrode with Kimwipes. Avoid rubbing the electrode while wiping.

3. Since pH reading/measurement is temperature dependent, allow the standard buffers (generally taken out from a refrigerator) to reach to the room temperature. Pour the buffer into a beaker for calibration.

4. Press the calibrate button to begin calibration. Place the electrode in the buffer for 7.0. Allow the pH reading to stabilize (by waiting for approximately 1 minute). Once the reading is stabilized, set the pH meter 7.0 by pressing the measure button. Preferably do not pour back the used buffer back into its original container.

5. Rinse and dry the electrode with a lint-free Kimwipe.

6. Place the electrode in the second buffer (pH 4.0) and begin reading by pressing the measure button. Once the reading has stabilized, set the pH meter to pH 4.0.
7. Rinse and dry the electrode with a lint-free Kimwipe.
8. If needed, calibrate the pH meter with third standard pH buffer (10.0).
9. Now, the pH meter is ready/calibrated for taking measurement of pH of unknown samples. Never use any brush or cloth to clean to electrode.

Measurement of pH of a Solution

1. Rinse the electrode with distilled water under and blot dry with Kimwipes Place the electrode in the sample, press the measure button and leave the electrode in the sample for approximately 1–2 minutes.
2. Once the reading has stabilized, press the measure button and note down the pH value of the sample.
3. Rinse the electrode with distilled water and blot dry with a lint-free Kimwipes. You can take reading of other samples or store the pH meter for using next time.
4. Keep the electrode submerged in 4M KCl solution when not in use for a longer time.

Care and Maintenance of pH Meter

- When pH meter is not in use, the pH electrode should not be kept dried (Figure 5.5). Instead, it should be kept wet to maintain workability of the electrode (pH meter). This will substantially increase the longevity of the electrode, save the time in preparing pH meter for measurement, and save money (needed to replace the electrode very frequently).

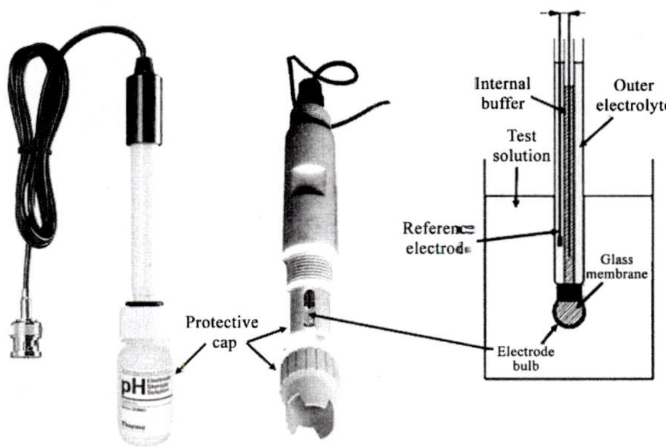

Fig. 5.5: pH electrode.

- Always keep the pH electrode moist. The best solution to keep the electrode wet (during short or long term storage) is by submerging it in 4M Potassium Chloride solution (4M KCl), which is also known as pH Storage Solution or Electrode Storage Solution. If KCl solution is not available (for any reason), use the pH 7 (or 4) buffer solution. KCl solution is generally available from many of the laboratory consumable suppliers. Fill the cap/container with enough 3M or 4M KCl solution to cover the glass bulb and maintain the level of solution to keep the bulb moist (Figure 5.6).

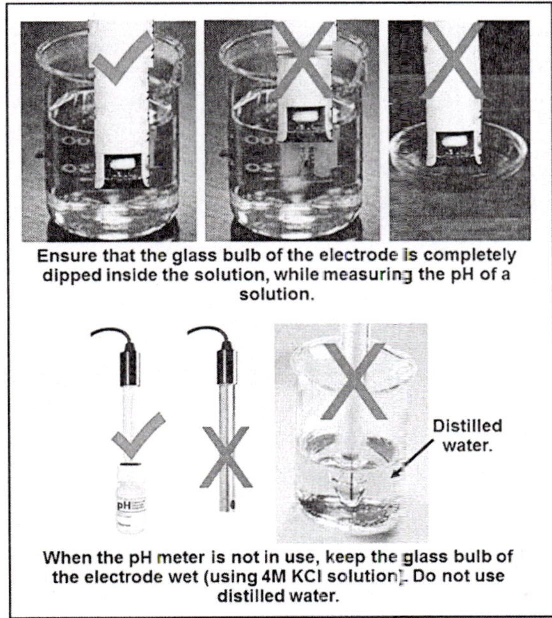

Fig. 5.6: pH meter care and maintenance.

- It is also very important that the pH electrode should not be stored in distilled/deionised water, as this (having little or no ions) may cause the ions (inside the electrode) to leach out of the glass bulb, which may render the electrode unserviceable.
- On storage, it may be noticed that KCl crystals are accumulated on (outside) the pH electrode. This is a normal phenomenon and these do not interfere with the working of the instrument. If this happens, gently wash the electrode with distilled water and blot dry the electrode before use. However, care should be taken not to put pressure on the electrode while blot drying/wiping.
- The pH electrode comes with a protective cap over the glass bulb to protect the electrode/bulb from any damage. Moreover, the electrode is protected with a rubber cap and kept submerged in a 4M KCl solution. Remove the rubber cap prior to using the electrode.
- Generally, the lifespan of a pH electrode is 12–18 months. However, it also depends on how frequently the pH meter is used, the temperature of operation, type of sample used for pH measurement, and the care/maintenance of the electrode.

Maintenance of Portable pH Meter

To ensure accurate readings and prolong lifespan of a portable pH meter (Figure 5.1), its proper maintenance is crucial. One of the most important aspects of its maintenance is regular cleaning. After each use, the meter should be wiped down with a soft cloth to remove any residue/debris. If needed, a mild detergent can also be used to clean the electrode. If necessary, use specifically designed cleaning solution for pH meters once a week to remove any deposits that might affect accuracy of the pH meter.

Proper storage is also important for maintaining the performance of the portable pH meter. When not in use, the electrode must be kept moist in a storage solution. The pH meter should be stored in a dry, cool place, away from direct sunlight and extreme temperature. It is also important to ensure that electrode of the pH meter does not come in contact of any hard surface or object that may damage the electrode.

Over time, the electrode may get damaged, leading to its inaccuracy. To ensure consistent performance, it may be necessary to replace the electrode. This can be done by carefully removing the old electrode and attaching a new one according to the manufacturer's instructions.

Laminar Flow

Laminar air-flow is an enclosed cabinet equipped with a HEPA (high-efficiency particulate air) filter for contamination-free airflow system. Thus, it provides a germ-free environment by unidirectional circulation of air at uniform velocity (without any turbulence) in the range of 0.3 – 0.5 m/s to conduct experiments.

Laminar Air-flow Hood

Clean bench and biosafety cabinet are common examples of laminar air-flow hoods. It is a laboratory enclosure, designed to carefully guide the air through a HEPA filter. Some of these hoods can protect objects placed on the work-surface from contamination. Others prevent the users from being exposed to the microbes being worked upon. The principles of laminar air-flow were first developed in the early 1960s. Over the generations, the way air travels within the enclosure of a cabinet was shaped, and it is still relevant today in modern laboratories. Many different types/categories of laminar air-flow hoods exist. All of these are based on unidirectional airflow to help maintain sterility, prevent cross-contamination, and reduce turbulence.

Principle of Laminar airflow chamber

The function of laminar flow is to remove all the contaminants in the air and maintains a sterile environment, according to the high-efficiency particulate air system. Elements essential to the operation of a laminar flow include: a high-efficiency particulate air filter, filter-pad, and a fan. The fan/blower sucks air that has been pre-filtered through filter pads to trap air pollutants. The pre-filtered air then passes through a HEPA filter to remove contaminants like spores, bacteria, and dust, resulting in sterile air-flow at uniform rate and single direction (Figure 5.7).

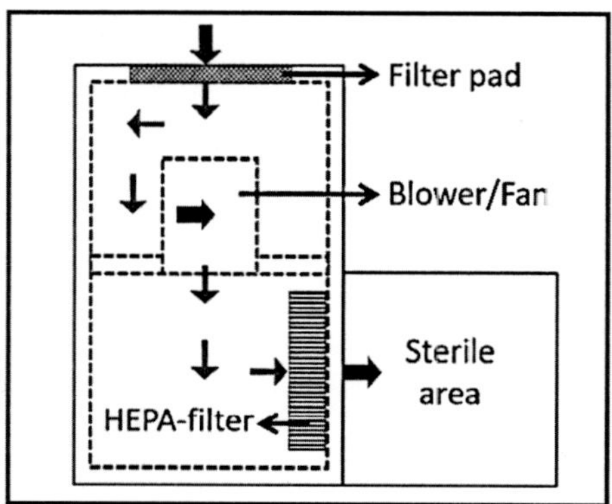

Fig. 5.7: Working principle of a laminar air-flow hood (*Source*: biologyreader.com).

Usage of Laminar Air-flow Hood

- Ideal for performing preparatory culture platings.
- Tissue culture can be performed in a horizontal laminar flow.
- Vertical laminar airflow offers the best performance against a variety of dangerous/contaminating specimens.
- Ideal for preventing particle-sensitive substances and semiconductor wafers from contamination.
- Culturing of biological specimens.
- Provides desirable environment for performing various tasks in the fields of medical, pharmaceutical, and electronic industries.

Types of Laminar Air-flow

a. **Vertical displacement air-flow chamber:** It is often used in laboratories. In this, air flows from the top of the laminar flow hood. Air enters through the HEPA filter (placed on the top) and blows from the top to the working area (Figure 5.8). As air does not directly blow to the user while operational, it is much safer than a horizontal laminar flow.

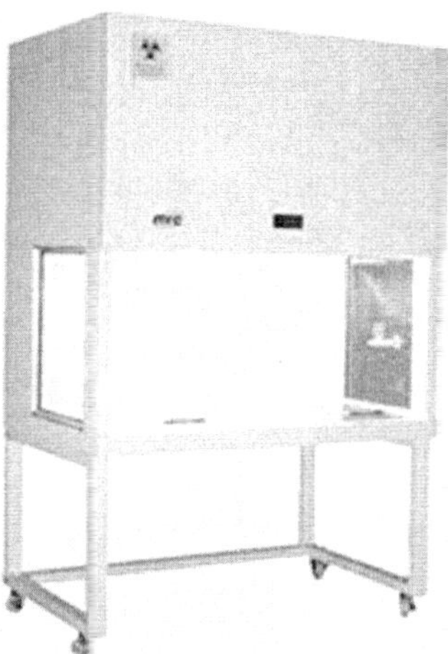

Fig. 5.8: Vertical laminar air-flow hood.

b. **Horizontal air-**flow chamber: In this, air enters the HEPA filter placed at the back of the hood. Air flows horizontally from the front. In the horizontal laminar air-flow cabinet, the air moves parallel to the work area, and the workstation is disinfected by a constant air-flow (Figure 5.9).

Components of a Laminar Air-flow Hood

a. **Enclosed cabinet:** This is made of stainless steel. The cabinet provides an insulated air-tight system. This protects the sterile air inside, from the outside environment. Users can be opened manually at the front and the height of opening can be adjusted, as desired.

Fig. 5.9: Horizontal laminar air-flow hood.

b. **Working platform:** It provides a space for conducting various experiments/tasks. It is also made of stainless steel. Petri dishes, blowers, biological samples and necessary materials can be stored herein.

c. **Filter pad:** This is the pre-filter or primary filter that inhales the air from the room. Filter pads are installed at the top of the vertical laminar flow cabinet while at the bottom of a horizontal laminar air-flow cabinet. It traps particulate matters (>5 microns in size) present in room/air.

d. **Blower or Fan:** It drags the pre-filtered air and throws it towards the HEPA filter. The blower in the vertical laminar flow is under the filter pad; whereas, the blower is placed next to the filter pad in the horizontal laminar flow.

e. **HEPA filter:** A final filter that provides 99.9% efficiency in removing 0.3 micron particles. HEPA filter is mainly responsible for creating a sterile condition.

f. **UV germicidal lamp:** To sterilize the work-place and the items (petri-plate, test tubes, culture medium, etc.) in the chamber prior to starting the laminar air-flow hood, UV lamp is put on at least 15 minutes before working the work.

g. **Fluorescent lamp:** It provides visible light in the working chamber during the use of laminar flow.

Procedures to be Followed While using a Laminar Flow Hood

1. The working area/surface must be wiped with 70% ethanol, all the materials lying in the working area should be exposed to UV light at least for 15 minutes.
2. The items sensitive to UV rays must be removed from the cabinet, before switching on the UV light.
3. The glass front door of the hood must be closed and UV light should be switched on for 15–30 minutes to ensure surface sterilization of the working bench as well as all the equipment kept inside the chamber.
4. UV light must be switched off and then the air-flow should be switched on.
5. The air-flow must be switched on before opening the lower-half of the front door and starting the work in the chamber.
6. The front door should be opened, fluorescent light/visible light should be on and the UV light must be off during working inside the chamber.
7. Hands must be well sterilized with 70% ethanol.
8. One must take care to keep the wet hands away from the flame, as ethanol id inflammable.
9. Forceps, inoculation loop, scalpels, etc. to be used in the chamber while performing the experiment must be flame sterilized and cooled down under the sterile condition of laminar air-flow hood to avoid contamination.
10. Once the work in the laminar flow is complete, clean the laminar work place, switch-off the light and air-flow, and immediately close the front door.

Care to be Taken While using Laminar Air-flow Cabinet

1. The UV light should be off while working inside the laminar flow, as it may cause damage to the worker as well as the material worked upon.
2. The waste material should be removed from the hood.
3. Hands should be cleaned/sterilized properly with 70% ethanol.
4. The worker must keep hands inside the cabinet, should not touch face, clothes, and if the hands come out of the hood, they should be sterilized again using 70% ethanol.
5. Large objects should not be kept in front of the working area, as it might create obstacle in air-flow and may suspend contaminants.
6. The flame of a burner should be adjusted and kept in front of the operation.
7. The work platform, glass shield and all other components present inside the cabinet should be sterilized before and after the completion of the work.
8. Operator should wear lab-coat, protective equipment (as per the requirement of the experiment) while working inside the laminar flow.

Not to do in a Laminar Air-flow

1. Don't put any waste/unnecessary item in hood of laminar airflow.
2. Don't wear any jewelry/threads on hands as it may carry contamination in the work area.
3. Don't block the space/air-flow between the filter and working area inside the laminar flow.
4. Don't place any large item in the back of the hood to avoid disruption in air-flow or any contaminant in the lower (air suction) area of the laminar hood.

Thermometer

Thermometer is a device that measures temperature or its gradient, which is measured by using a material that expands when heated or comes in and response to the change in temperature. This causes the length of liquid (Mercury) column to become longer or shorter depending on the temperature. A thermometer has two important elements: the temperature sensor (e.g. a bulb in case of mercury thermometer) in which some physical change occurs due to changes in temperature, a means to convert/visualize this physical change into a value (e.g. scale on a mercury thermometer). There are different types of thermometer (Figure 5.10) according to their usage and manufacturing (Figure 5.11).

Mercury is a liquid at room temperature and it has the highest coefficient of expansion. Therefore, even a slightest change in temperature can be noticed. Also, the very high boiling point (357 °C) of mercury makes it suitable for measuring higher temperature. Another reason why mercury is used in thermometers is that it does not stick to the surface of the glass and has a shiny appearance. It shows a linear expansion ability that provides a basis for precise measurements. More importantly, it is a good conductor of heat, having quick response time, and has a wide temperature range for response.

Formula/Conversion Factor

- Fahrenheit (°F) = [9/5 × °C] + 32
- Celsius (°C) = 5/9 (°F – 32)
- Kelvin (°K) = 273 + °C

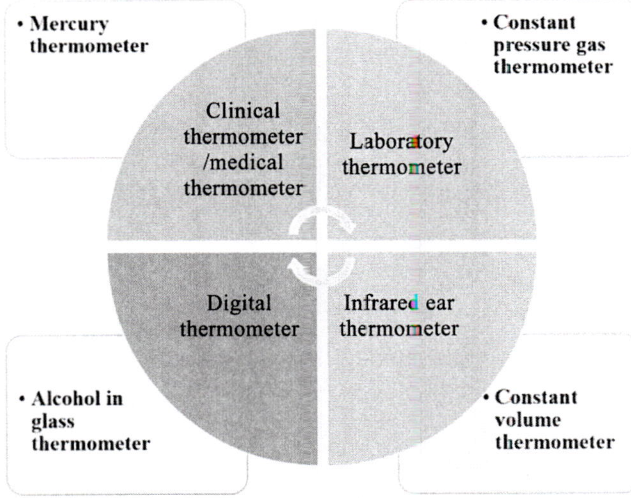

Fig. 5.10: Types of thermometer

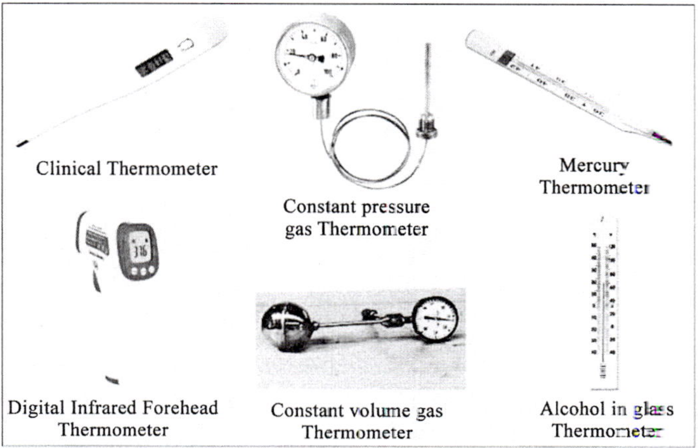

Fig. 5.11: Different types of thermometer

Things to Remember while Using Thermometer

- A thermometer is a device that is used to measure temperature. It usually comprises a small glass tube with a thin liquid column that raises/lowers in response to the temperature changes.
- Thermometers are frequently used to monitor the changes in a laboratory/ industry.
- To monitor body temperature, a clinical/ medical thermometer should be used. However, while using a mercury clinical thermometer, ensure to bring down the mercury (back into the bulb) by down-ward jerk (keeping the mercury bulb of the thermometer at lower side) before reusing it. However, this is not needed while using a digital clinical thermometer or a mercury laboratory thermometer.
- Laboratory thermometer is used to determine the room temperature or heated solid/liquid.
- Infrared thermometers can be used to take reading/temperature from a distance (without touching the body/material.

Precautions While using a Thermometer

- Clean the thermometer before use.
- Dip the clinical thermometer into an antiseptic liquid before and after use.
- It should be handled with care, as a thermometer may break easily and the mercury inside (a toxic substance) may come out.

- Avoid touching the bulb of the thermometer either before, after or during measuring the temperature.
- While measuring body temperature using a digital thermometer, keep the thermometer in contact of the body until (1–2 minutes) it makes a beep sound.

Viscometer

Viscometer is used to measure the viscosity and flow of fluid/liquid (Figure 5.12). Viscosity is the measure of how resistant a liquid is to motion when force is applied to it. In a viscometer, either the fluid remains stationary and an object moves through it, or the object is stationary and the fluid movesthrough it. Viscosity is important for the lubricant used in machines. If it is too less, it will be less efficient and more wear of the internal components of the machine. If it is too high, it makes the machine work harder.

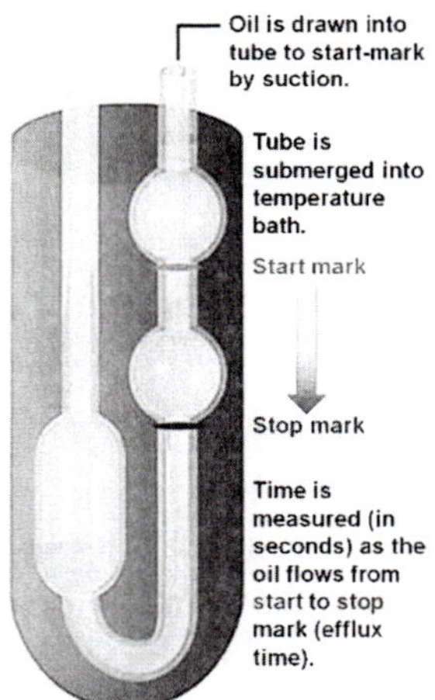

Fig. 5.12: Capillary viscometer

There are six different kinds of viscometer

1. Capillary viscometer,
2. Orifice viscometer,
3. Rotational viscometer,
4. Vibrational viscometer,
5. Ultrasonic Viscometer,
6. Falling ball viscometer (Figure 5.13).

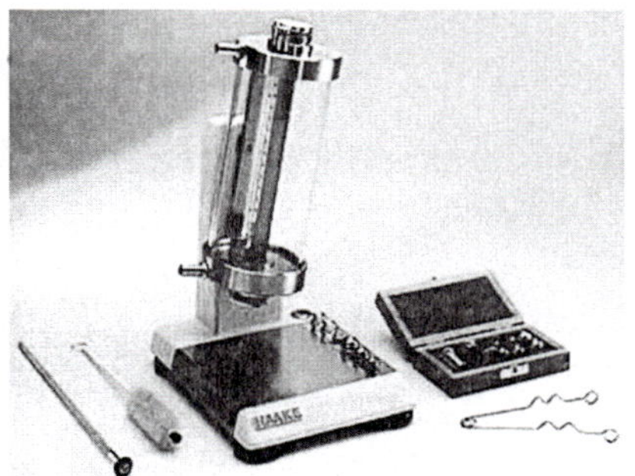

Fig. 5.13: Falling ball viscometer.

Handling of viscometer

- The viscometer tube must be dry before using. The use of dry, dust-free air or nitrogen is recommended.
- The residue of previous sample coating the inner side of the viscometer can be removed using chromic acid cleaning solution or a non-chromium-containing, strongly oxidizing cleaning solution.
- The viscometer should be held in a vertical position during the flow of the liquid.
- When placed in the temperature bath, the tube should be in thevertical position, free from vibration, and immerged at the specified depth.
- While sucking the liquid, no air bubble should be formed/entered inside the capillary.
- Manufacturer's instructions should be followed.

Magnetic Stirrer

A magnetic stirrer is a device used in laboratories to mix the solution continuously. It consists of a rotating magnet inside the unit and requires a magnetic bead to be placed in the solution to be stirred (Figure 5.14).

Principle

An external magnetic field is applied to the magnetic bead/bar placed in solution, which rotates with the rotating magnet of the instrument. The speed of rotation of the magnet inside the instrument is increased or decreased (using the knob) to increase or decrease the speed of rotating/stirring solution in the container/beaker.

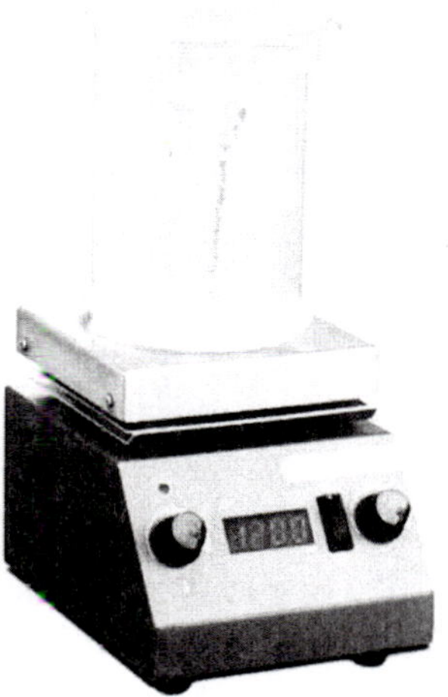

Fig. 5.14: Magnetic stirrer.

Usages of magnetic stirrer

- This is very useful when one needs to mix/dissolve a liquid or solid component to make a homogeneous mixture/solution.
- This is mainly used to stir/agitate a liquid for speeding up the mixing/dissolving.

- This minimizes the risk of contamination since a magnetic bead/bar coated with inert material (Figure 5.15) is placed inside the solution to be stirred.
- Magnetic stirrer can only be used for relatively small volume (<4 liters) of solution.
- For larger volume, mechanical stirring (an overhead stirrer) is used.
- A magnetic stirrer can also be used to accelerate the rate of reaction and the evaporation rate.

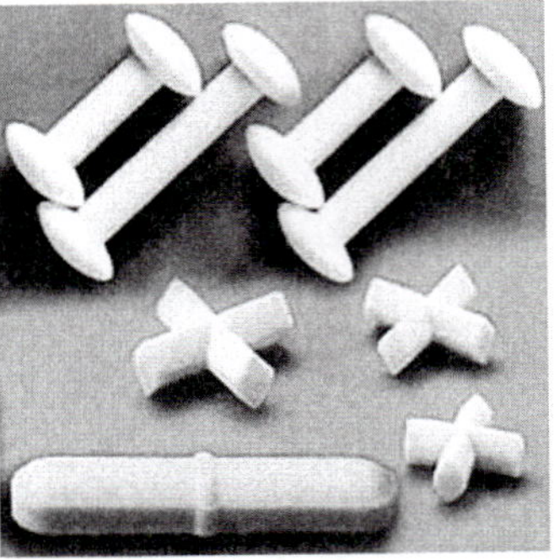

Fig. 5.15: Magnetic beads/bars of different types.

Handling of Magnetic Stirrer

1. Place the container on the magnetic stirrer platform.
2. Put the clean magnetic bead into the solution and let it settle.
3. Turn on the magnetic stirrer by switching on the power switch, and then rotate the speed knob in a clockwise direction to increase the speed slowly.
4. Speed of rotation can be controlled using the knob as per the requirement. The magnetic stirrer can be used with or without hot plate.
5. To turn-off the magnetic stirrer, rotate the knob counter-clockwise until a click sound is heard, and the stirring/rotation of magnetic bead stops. Switch-off the power knobs.
5. To remove the magnetic bead, a clean magnetic bead retriever is put inside the solution where the magnetic bead stick to the submerged tip (magnetic portion) of the retriever (Figure 5.16) and the bead is taken out of the solution along with the magnetic bead retriever. The magnetic bead and the retriever is washed/cleaned and kept/stored for next round of use.

Warning: *Avoid using the magnetic stirrer continuously at very high speed for longer time (>4 hours). If the instrument becomes hot, let it cool down to room temperature and then reuse it.*

Since magnets can lose their properties if they are not stored properly, the bar magnets should be stored in pairs with their unlike poles on the same side

(Figure 5.17) separated by a piece of non-magnetic material (e.g. wood) and covered by magnetic materials (e.g. soft iron) at the edges/poles.

Microwave-oven

Microwave oven is a common appliance, particularly in food industry. It is used for heating/drying a sample.in a quick manner (Figure 5.18).

Fig. 5.17: Storage of magnetic beads.

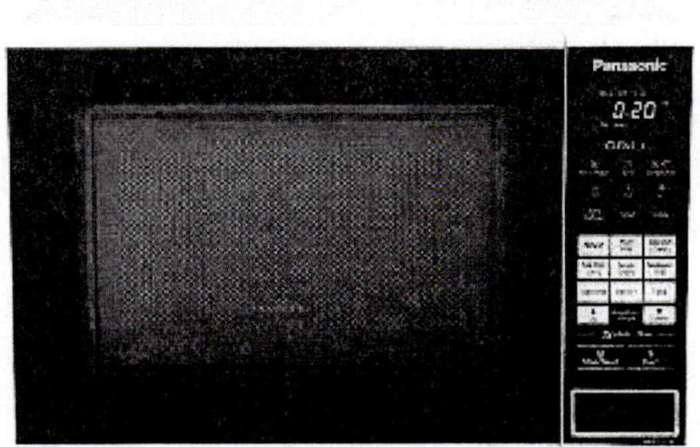

Fig. 5.18: Microwave oven.

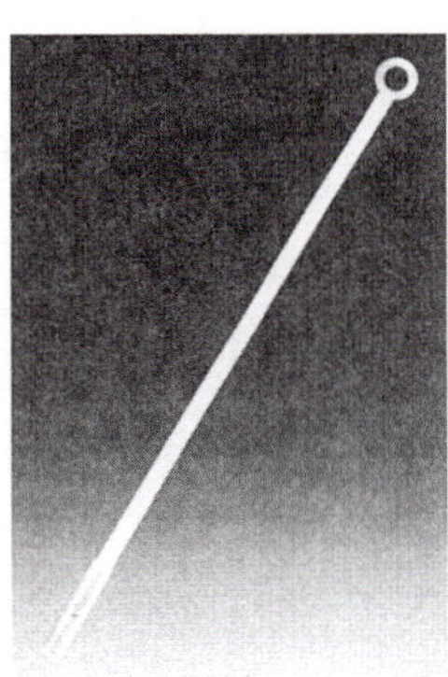

Fig. 5.16: Magnetic bead retriever.

Working Principle of Microwave-oven

- Microwave energy is electromagnetic radiation with the frequency between 3 MHz – 30000 GHs. The two major mechanisms are dipole-rotation and ionic-polarization. When a moist sample is exposed to microwave radiation, molecules carrying dipolar electrical charges (such as water) rotate as they attempt to align their dipoles with the rapidly changing electric field. The resultant friction creates heat, which is transferred to the neighbouring molecules. Thus, microwave heat the material rapidly and thoroughly, even a thicker one. However, a microwave selectively heats the areas having higher water content. Internal temperature of a moist microwave-heated sample may reach to the boiling point of water. At this point, free moisture evaporates inside the product, which causes a vapor pressure gradient that expels moisture from the sample. The internal temperature remains at the boiling point until all free moisture is evaporated, following which the internal temperature rapidly increase. Very rapid drying occurs without overheating the surface of the material, because heat is generated internally.

- Microwave oven usually used in a laboratory uses the frequency of 2450 MHz. The oven come equipped with an internal electronic balance and usually allow fine-tuning of the microwave power.

Handling of Microwave-oven

Hazards

The use of microwave ovens for simple heating or defrosting in a laboratory may pose the following hazards:

- Exposure to microwave radiation from a faulty/modified unit.
- Electric shock from ungrounded/faulty unit.
- Ignition of material being heated.
- Pressure build-up in sealed container.
- Sudden boiling of liquid in an open container even after removal from the oven.
- Contamination of food product with chemical residues, particularly in laboratories.

Risk management & control measures

To minimize the risk of above-mentioned hazards, appropriate control measures (Dos and Don'ts) should be adopted to minimize the risk to health and safety from the microwave.

Dono'ts

- Never heat flammable liquids/solids, hazardous substances or radioactive materials in a microwave oven.
- Never use a microwave oven placed in a laboratory for heating food/ edible materials.
- Avoid using a heat sealed containers in a microwave oven.
- Avoid using a bottle with a restricted neck opening in a microwave oven.
- Never use/place a metallic object in a microwave oven.
- Avoid overheating a liquid in a microwave oven.
- Avoid overloading a microwave oven.

Do's

- Ensure that the oven is adequately ventilated.
- Conduct regular inspections to ensure that the sealing surfaces are clean and do not show any sign of damage.
- Ensure that microwave oven is electrically grounded and connected using a properly rated three-wired cord, pin, and plug.
- Always use microwave-grade plastic vessels with a pressure relief valve.
- Use appropriate protective equipment (gloves) when removing heated materials from microwave oven.

Incubator

Incubator is an insulated and enclosed device that can provide optimal conditions with respect to temperature, humidity, CO_2, O_2 content, pressure, and other environmental factors required for the growth of an organism. A laboratory incubator is usually used to grow and maintain microbiological or cell cultures under optimal atmospheric conditions. It can be used for the growth of unicellular as well as multicellular organisms by providing controlled conditions for cell/tissue culture.

Principle of incubator

Incubators work on the principle of maintaining proper atmospheric conditions (Figure 5.19) for optimum growth of microorganisms.

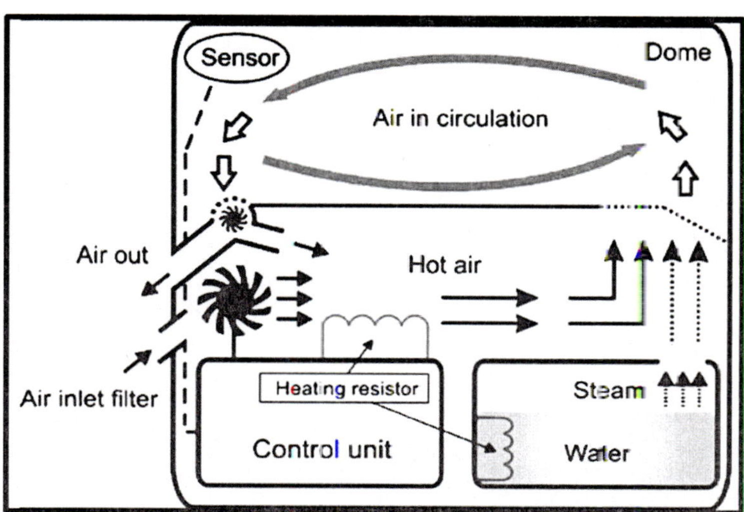

Fig. 5.19: Working of an incubator

It may have heating and cooling systems that help in adjusting temperature inside the incubator according to the organism and the need. A thermostat fitted on the incubator maintains a constant temperature that can be observed on a display. Similarly, it can also maintain the concentration of O_2 (BOD incubator) or CO_2 (CO_2 incubator) and humidity required for the growth of organism. Shaking at different/constant rotation per minute (RPM) can also be maintained (incubator shaker) that allows continuous shaking of the culture required aeration to cells (Figure 5.20).

BOD incubator

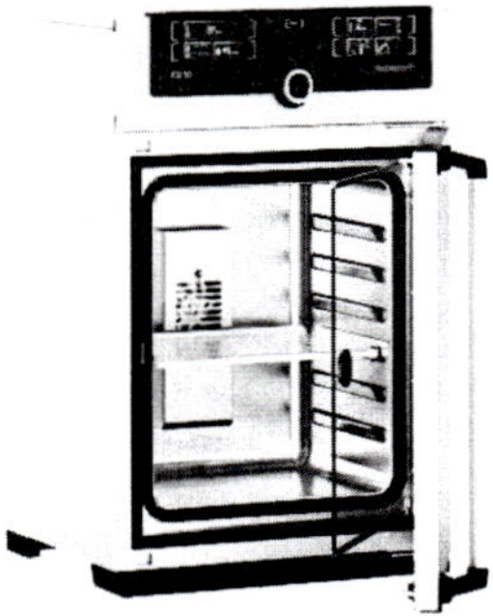

CO_2 incubator

Fig. 5.20: Different types of incubator

Parts of an incubator

Cabinet: Main body of the incubator made of double-walled cuboidal enclosure with space between the two walls filled with glass wool to provide insulation to the incubator.

- **Door**: To close the insulated cabinet.
- **Control panel**: Switches and indicators that allow the parameters to be set for controlled conditions.
- **Thermostat**: To regulate the desired temperature inside the incubator.
- **Perforated shelves**: Trays that provide space for placing materials and allow free movement of air inside the incubator.
- **Asbestos door gasket**: Provides airtight sealing between the door and the cabinet.
- **Thermometer**: Helps reading temperature inside the incubator.
- **Humidity and gas control**: Controls the humidity concentration of gas inside the incubator.

Running an incubator

1. Before using an incubator, it should be ensured that there is no previous culture/contamination present in the incubators. The same incubator can be used for growing multiple organisms/cultures, if they require the same set of parameters.
2. The door of incubator is closed and the incubator is switched-on. After some time of heating the temperature inside the incubator should be checked using a thermometer or the display on the incubator.
3. If the microbe requires a particular concentration of gas (O_2 or CO_2) and/or humidity, the desired parameters should also be set and checked.
4. Once the all the set parameters have been achieved, place the materials on the shelves and incubate for the desired time period.

Usage of incubator

- Laboratory incubator provides a controlled, contaminant-free environment for safe, reliable with cell and tissue cultures by regulating culture conditions like temperature, humidity, and CO_2.

- Microbiological incubator is used to grow microbial cultures.
- Incubators can also be used to maintain the culture to be used later.
- Some incubators are used to increase the growth rate of organisms, having a prolonged growth rate in the natural environment.
- Specific incubators are used for the reproduction of microbial colonies and subsequent determination of biochemical oxygen demand (BOD) (e.g. Waste Water Monitoring).
- Incubators are also used for breeding insects and hatching eggs in Entomology.
- Incubators are used to create controlled conditions for storage of sample before they can be processed in a laboratory.

Precautions while using incubator

- Avoid opening the door of incubator frequently, which may lead to fluctuation in temperature of the cabinet, as microorganisms are sensitive to changes in temperature.
- The cultures should be placed inside the cabinet once the desired parameters for growth of the organism have achieved.
- The microbial culture plates should be placed upside down (lid at the bottom) to avoid condensation of water vapour onto the media and disturbance in microbial colony development.

Water-bath

Water bath is a laboratory instrument which is used to heat/incubate samples by dipping in hot water. The temperature of water is maintained constant at desired level to incubate the samples over a period of time. It can maintain temperature up to 100 °C. The main parts of a water bath are (i) tank (made up of steel), (ii) heater, (iii) thermometer, and (iv) regulator.

- It is useful when several tubes are to be incubated at a time.
- It is used for warming of reagent, melting of substance (slowly, without direct heating) and incubation of cell culture.
- It is also used to enable certain chemical reaction to take place at higher temperature.
- It is a preferred way of heating a flammable chemical, instead of open flame heating, to avoid ignition.

Type of water bath

There are three types of water bath: (i) Water bath with lead, (ii) circulating water bath, (iii) shaking water bath (Figure 5.21).

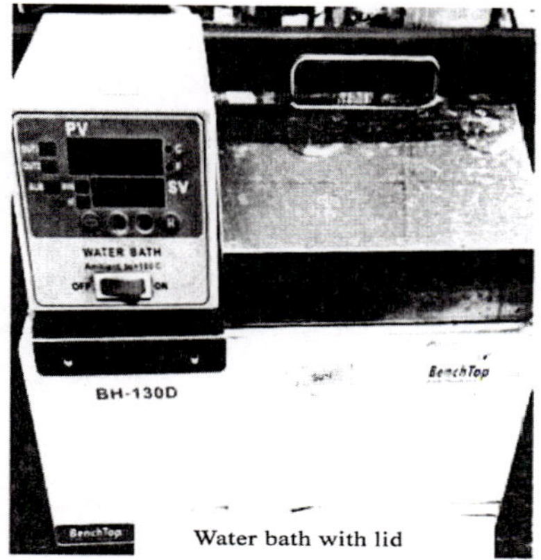

Water bath with lid

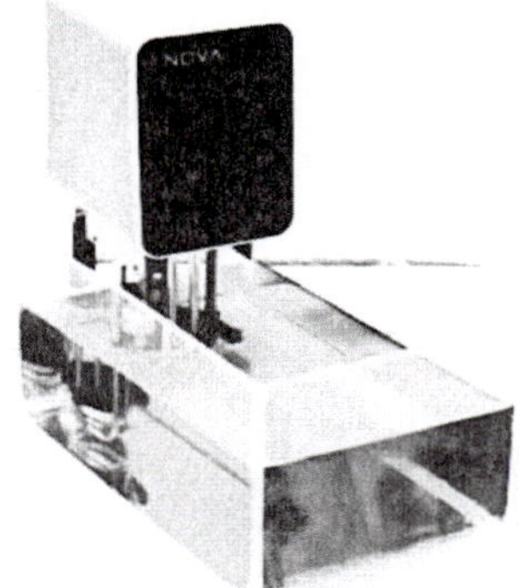

Circulating water bath

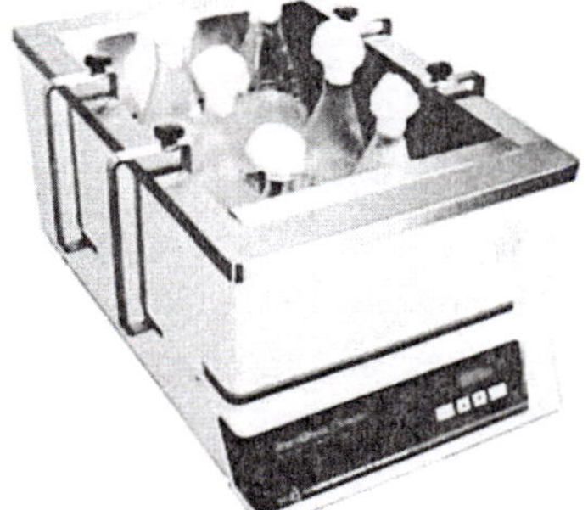

Shaking water bath

Fig. 5.21: Different types of water bath.

Handling of Water Bath

Keep the water bath on a stable/solid surface.

- Do not use tap water in water bath as it contains lot of ions/salts, which accelerate the rate of corrosion of the tank of water bath.
- The level of water should be above the level of immersion heater to avoid overheating of the heater.
- Empty the water tank and clean/dry it to prevent rusting. Rusting in water tank can cause contamination of samples.
- To check proper working of water bath, use a thermometer to measure temperature of water in the tank and then compare the observed temperature with t value with the set and displayed value.

Precautions while using water bath

- Water level should be checked regularly and filled with distilled water. This is necessary to prevent deposition of salts on the heater.
- Disinfectants can be added to prevent growth of microorganisms or raise the temperature to >90 °C once in a week for half an hour to decontaminate the water/tank.
- The water bath should be covered with its lid to prevent evaporation, particularly at higher temperature.

Sand-bath

Sand-bath is ancient laboratory equipment, used by the alchemists. It is made of a container filled with sand, fitted with heating devised, and heat-controller to regulate the temperature. A sand bath is used to heat the material to a higher temperature (>250 °C). Sand takes a long time to heat up and a longer time to cool down.

In a sand-bath, a heatproof dish contains clean sand to a depth of 25 mm. A heating devise like stove or electric heater is used. Temperature controlled is also used to regulate the temperature (Figure 5.22). The reaction vessel is kept partially covered by sand and the sand-bath is switched on. Heating duration can also be set if timer is fitted on the sand-bath.

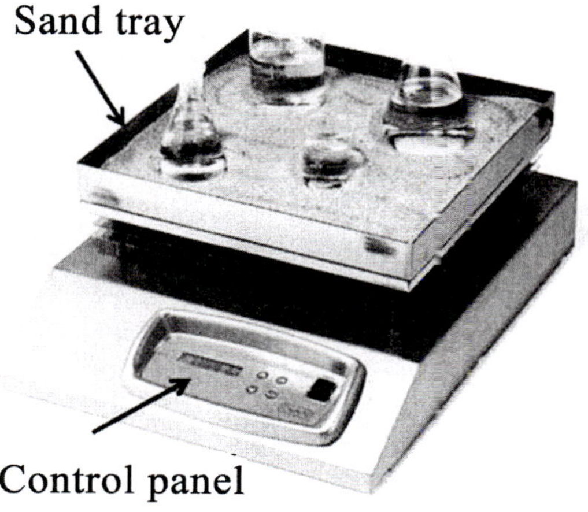

Fig. 5.22: Sand-bath.

Advantages

Sand-bath can be used to heat solutions to very high temperatures (>250 °C) with uniform heating. Because sand is a poor thermal conductivity and dissipates heat faster, the bottom sand layer should be thinner to facilitate fast heating. However, thermal efficiency of sand-bath is lower compared to other methods of heating. One should use heat-protecting (Asbestos) gloves while using a sang-bath.

Oil-bath

Oil-bath is a laboratory heating device in which a high temperature, heat conducting oil (instead of water) is used. The oil is placed in a container which is heated using a heater and heat-controlling devise (Figure 5.23).

- Oil-bath is commonly used to provide higher temperature at steady rate, which is important for thermodynamic control of many reactions.
- Generally, mineral oil, silicone oil or industrial oil are used in an oil-bath. The oil should be preferably of low cost, it should be cheaper, and easily available.
- Specific oil is used for a specific purpose/experiment based the temperature requirement. Since different oils have different boiling points, it is desirable to select an oil with a boiling point as close as possible to the required temperature for the experiment.

- At the standard atmospheric pressures, the maximum temperature which can be achieved using a water bath is 100 °C. While using an oil-bath the temperature can be achieved up to the temperature about 250 °C.
- Operation of oil-bath is same as that of a water-bath, as in both the cases the experimental material is heated by dipping in oil/water.
- For the heating at less than 250 °C, it is better to use oil-bath than a sand-bath. An oil bath has more accuracy and thermal efficiency than a sand-bath.
- Now-a-days, oil is electrically heated through an immersion coil which is connected with a temperature controller to adjust/regulate the temperature of oil.

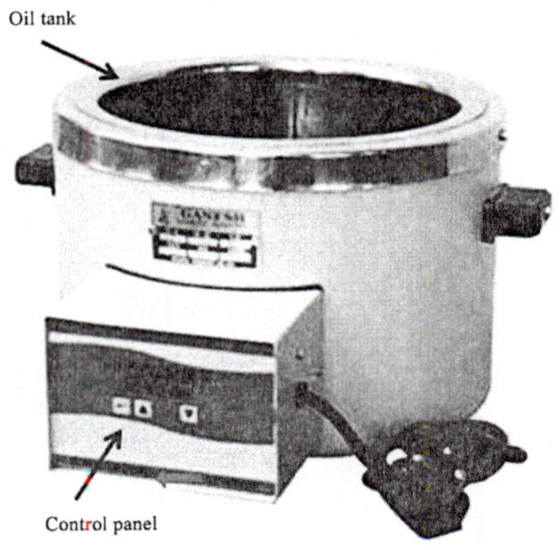

Figure 5.23: Oil-bath.

Safety measures for using an oil-bath

- Hot oil can be more hazardous for the user; hence one should be more careful while using it.
- Splash of oil from oil-bathe may cause severe burn injury.
- Glassware immersed in oil is slippery. If it is not handled carefully, there is greater chance of dropping (breakage) and loss of the reaction material.
- Oil-bath must be labelled with the oil in it and a safe temperature range it should be used.
- Overheating of oil should be avoided since oils are highly susceptible to ignition. If smoke starts coming out of the oil, turn it off and use at lower temperature.

- Generally, oil expands in volume upon heating, hence overfilling of the oil tank should be avoided.
- It is always advisable to keep water away from the hot oil, as admixture of hot oil and water cause splattering of oil leading to oil burn.

Let's Check Your Understanding

1. Which type of substance can be heated using water bath?
2. What is the function of a thermostat?
3. What is the maximum temperature up to which an oil-bath should be used?
4. Given below are the two statements of commonly used baths in laboratory?

Statement I: The bath is often slightly hotter than the liquid inside the flask

Statement II: The liquid in the flask is often slightly hotter than the bath. In the light of the above two statements, choose the most appropriate answer from the options given below:

a. Both statement I and statement II are true

b. Both statement I and statement II are false

c. Statement I is correct but statement II is false

d. Statement I is incorrect but statement II is true

5. With regard to a water-bath, which of the following statement(s) are correct?

a. Shaking water bath is mostly used for serological and biological work.

b. Circulating water-bath is mostly used for serological and biological work.

c. The most common heating bath used in laboratories is a water-bath.

d. Non-circulating water-bath is also known as stirring water-bath.

e. Non-circulating water-bath relies primarily on principle of convection.

Choose the correct answer from the options given below:

a. A, B, C only   b. B, C, D only

c. C, D, E only   d. A, C, E only

6. What is the maximum temperature that can be achieved by an oil-bath at standard atmospheric pressures?

a. 100 °C   b. 200 °C

c. 250 °C   d. 470 °C

7. A magnetic stirrer is a device that employs _______ ________ to a magnetic bead (immersed in a solution) to spin/stir the solution.

   a. Electric field
   b. Magnetic field
   c. Thermal field
   d. Gravitational field

8. Magnetic bead is typically coated with _____________, which is chemically inert, not contaminating or reacting with the solution/reaction mixture.

   a. Aluminum
   b. Polytetrafluoroethylene
   c. Iron
   d. Hexafluoropropylene

9. The stirring speed of a magnetic stirrer is maintained between?

   a. 250–1500 rpm
   b. 5000–10,000 rpm
   c. 120–1400 rpm
   d. 1200–3000 rpm

10. The electromagnetic frequency used in household or laboratory microwave oven is usually

    a. 1450 MHz
    b. 2450 MHz
    c. 3000 MHz
    d. None of the above

11. Basic thermometer was invented by

    a. Galileo
    b. Newton
    c. Robert hook
    d. Flemming

12. Which type of Viscometer is used in oil industry?

    a. Capillary viscometers
    b. Orifice viscometers
    c. Rotational viscometers
    d. Vibrational viscometers

13. Part of an incubator that is used to set the desired temperature is

    a. Thermostat
    b. Cabinet
    c. Thermometer
    d. Control panel

14. CO_2 incubator is used for the growth of microorganisms that require CO_2 in the range of:

    a. 10–20%
    b. 0–5 %
    c. 5–10 %
    d. 20–25 %

15. The pH inside an incubator must be maintained.

True / False

Answer

1. Highly flammable.
2. Helps maintaining constant temperature.
3. Up to 200 °C.
4. c
5. d
6. c
7. b
8. b
9. c
10. b
11. a
12. b
13. a
14. c
15. False

6

Use and Handling of Vacupette and Micropipette

Since mouth pipetting of liquid reagents in a laboratory is not advisable, different tools are needed for safe, easier, and precise dispensing of liquid reagents. This chapter is devoted to make you aware of the tools like vacupette and micropipette which can be used for the handling of liquid reagents in a safe and precise way.

Vacupette

Glass pipettes have been traditionally used by sucking liquid solvent/solution directly using mouth. For safety reasons, mouth pipetting is no longer permitted in a laboratory. Hence, to fill a pipette vacupette is used by placing it on the top of glass or plastic pipette. It acts as a device to create vacuum for filling reagent in the pipette as well as for releasing the reagent with mechanical regulation on the flow out of the liquid. Vacupette can be of different types.

1. Rubber Bulb Vacupette

Rubber bulb vacupette is often used to transfer (pump up and discharge) liquid using a pipette when a larger quantity of chemicals is to be transferred. Small/tiny rubber bulb is attached to glass/plastic pipette for drawing little amount of reagents. Small vacupette is used to precisely discharge smaller volume/ drops.

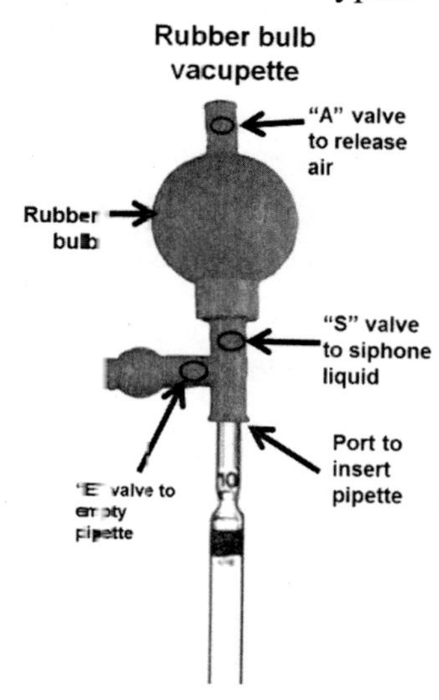

Fig. 6.1: Rubber bulb vacupette: their different parts.

Handling of Vacupette

- Insert the top of a pipette in the bottom of the rubber bulb vacupette (Figure 6.1).
- Remove air from the pipette by squeezing valve "**A**" at the top of the

vacupette while simultaneously squeezing the rubber bulb. The amount of air released depends on the size of the rubber bulb.

- Insert the tip of the pipette into the liquid to be dispensed.
- Siphon the liquid into the pipette to the desired level by squeezing valve "**S**". This uses the vacuum created in the bulb to draw liquid into the pipette. Be careful not to allow drawing liquid into the rubber bulb.
- Discharge the liquid from the pipette by squeezing valve "**E**" on the side tube. This allows releasing liquid at the desired rate and volume.

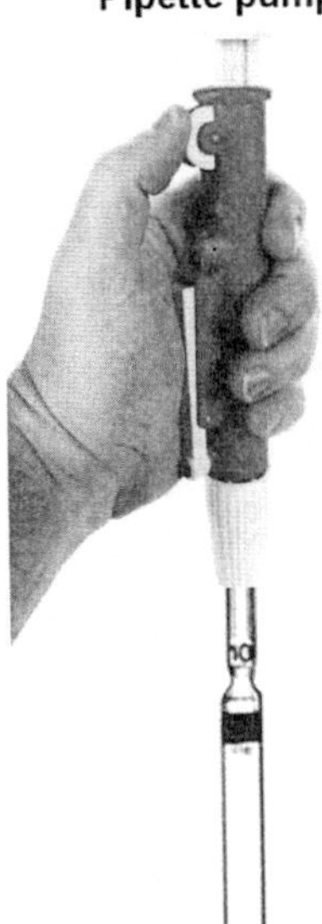

Fig. 6.2: Pipette pump.

2. Pipette Pump

Pipette pump is used to replace the rubber bulb. It contains a piston operated with a wheel or pipette filler with a flip (Figure 6.2). The flip design pipette filler include two valves and a top valve that may be removed for easy cleaning and one-handed use. Wheel design pump does not require squeezing of rubber bulb. The liquid is drawn up by spinning of the wheel and then released freely using the lever. This is quite easy/simple to use/clean.

3. Electrical Pipette Pump

Use of electrical pipette pump makes the sucking/releasing of liquid very easy and sophisticated. An electrical pipette pump operates on electricity and possesses two buttons to operate it (Figure 6.3). One of the buttons is used to fill the liquid in the pipette by pressing the button. The second button is used to release the liquid from the pipette in precise/desired volume.

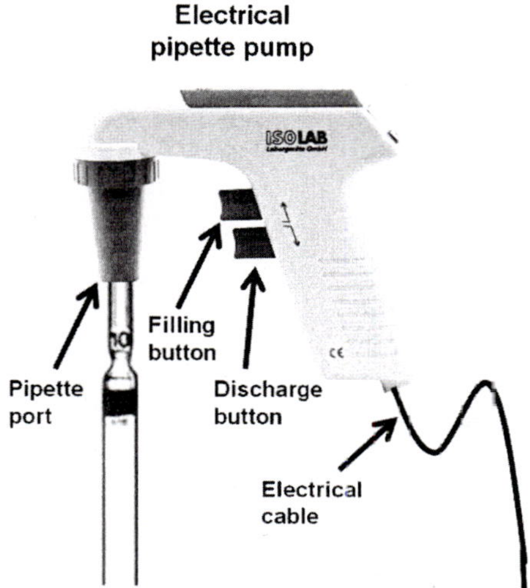

Fig. 6.3: Electrical pipette pump.

Precautions While using Vacupette

- Ensure that the pipette fits well into vacupette without applying much pressure. Do not force vacupette onto the pipette.
- Make sure the pipette tip is dipped into the liquid before sucking/ siphoning.
- Do not use a broken pipette, as end of a damaged pipette might be sharp enough to harm the vacupette.
- If a pipette pump is not sucking liquid inside, give a small twist and try pushing it a little harder. But if it still does not work, pipette pump might me broken; hence, replace it.

Micropipettes

Micropipette is used to measure very small to small amount of liquid with accuracy. It is used for molecular biology work to measure and transfer liquid reagent in microliter volume. It can be a single channel/tip or multi-channel pipette. Multi-channel pipette can be of 8 or 12 channel/tips.

Parts/Operation of Micropipette

a. ***Volume window:*** Volume of the liquid to be measured/dispensed using a micropipette can be seen/adjusted in the volume window. Desired value/ volume of liquid can be set within the range of the pipette, as different micropipette allows a different range of measurement volume.

b. ***Volume adjustment:*** Generally at the top of the micropipette, a knob is provided, which is rotated to decrease/increase the volume to be measured.

c. ***Pipette barrel:*** It is a tube filled with air containing a plunger to force out the set volume of air. On releasing the plunger, the set volume of liquid comes into the micropipette tip. There are two steps/stops in the plunger movement, which are used differentially for sucking up and releasing the liquid while pipetting.

d. ***Plunger:*** It is the most important part of a micropipette, which is used to suck and release the desired/measured volume of liquid.

e. ***Pipette tip:*** It is attached at the end of the micropipette for sucking in the liquid and then releasing the measured liquid from one to another container. Tips of different sizes are used for different micropipettes for measuring different volumes of liquid.

f. ***Eject button:*** An eject/push button is provided on the micropipette to eject the tip from the micropipette without touching the tip.

Commonly used micropipettes

Size	Range
P 2	0.2–2 μL
P 10	1–10 μL
P 20	2–20 μL
P 100	10–100 μL
P 200	20–200 μL
P 1000	100–1000 μL
P 5000	1000–5000 μL

Pipetting Liquid using a Micropipette

1. Micropipette plunger has 3 positions:
   - *Rest position*: Whenever a micropipette is not in use (stored after its use), its plunger should be kept at this position. Generally, the reading on the volume window should be 5 point lower than its maximum capacity.

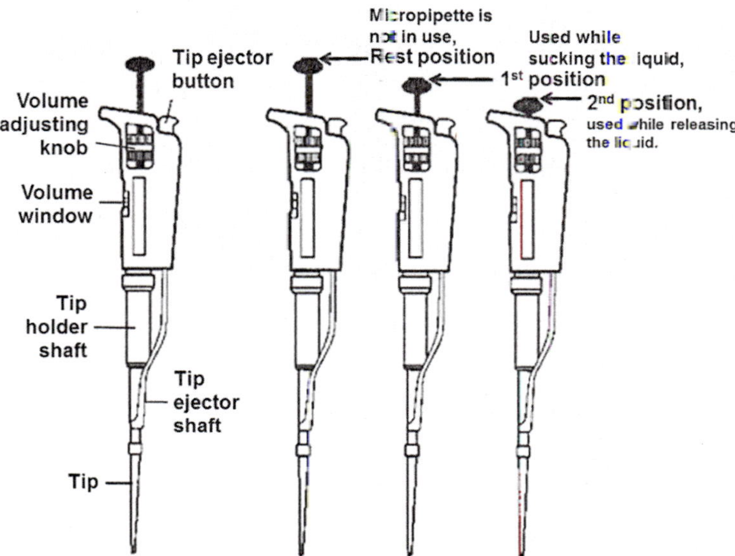

Fig. 6.4: Micropipette: parts and its handling.

- For example, a 2–200 μL micropipette should have been set at 195 value on its volume window while stored (not in use). This keeps the spring, inside the micropipette, in relaxed position for longer serving life.
- *First stop*: While using a micropipette for dispensing a desired volume of liquid, the plunger (many a time volume adjusting knob also) needs to be pressed until this step and dip into the liquid before releasing it slowly to suck the liquid.
- *Second stop*: When the sucked liquid needs to be released out of the micropipette completely (for precision in measurement), the plunger needs to be pressed until this step (last stop, beyond which the plunger cannot be pressed). This ensures complete ejection of the measured liquid out of the micropipette tip.

2. Fit the tip to the end of the shaft. Press down and twist slightly to ensure an airtight seal.
3. Hold the pipette in a vertical position. Depress the plunger to the first stop. Air equal to the volume of the setting (e.g. 100 μL) is displaced.
4. Immerse the tip into the liquid. Release the plunger back to the rest position. Wait a second for liquid to be sucked up into the tip. The volume of liquid in the tip will equal the volume of the setting of the micropipette.

5. Place the tip at an angle (10° to 45°) against the wall of the vessel receiving the liquid, for example a well of a microcentrifuge tube. Depress the plunger to the first stop, wait one second, press the plunger to the second stop to expel all the liquid

6. Move the end of the tip away from the liquid. Release the plunger to the rest position.

Precautions While Handling the Micropipettes

Manufacturer's instructions are supplied with all micropipettes. Please read and make sure you understand how to operate and care for the pipette.

- Treat micropipettes very gently as they are precision instruments.
- Keep upright when in use to prevent liquids running inside the shaft of the pipette.
- Do not leave pipettes lying on the workbench where they can be knocked off and damaged.
- Before use, make sure the volume has been correctly set. Adjust the volume before use. Most pipettes have a digital display of volume. Some brands have a micrometer setting, which can be difficult to read.
- Check all tips are securely fitted to pipette.

Air Displacement Pipette

An air displacement pipette works by moving air along with the movement of piston up and down, which creates a negative or positive pressure. Different parts of a pipette have been displayed in Figure 6.5). This causes the liquid to be aspirated into or expelled through the detachable plastic tip attached to it.

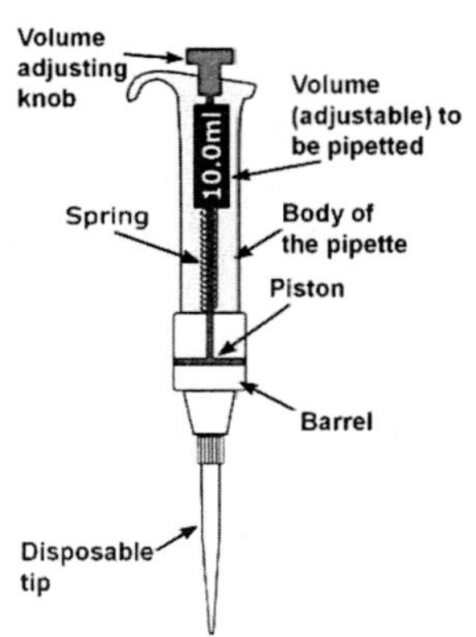

Fig. 6.5: Different parts of a variable volume pipette.

(i) ***Single channel pipette:*** Due to their versatility, accuracy, and wide volume range, variable volume air displacement single-channel pipettes are the most commonly used laboratory instruments. The same pipette can be repeatedly used for multiple applications. For use under sterile/contamination-less conditions, a plastic tip of varying/fitting size (either supplied as sterile pack or sterilized in the laboratory) is used (Figure 6.6). To avoid any contamination (to the pipette or from the pipette) aerosol tips (sterile or non-sterile) are used (Figure 6.7).

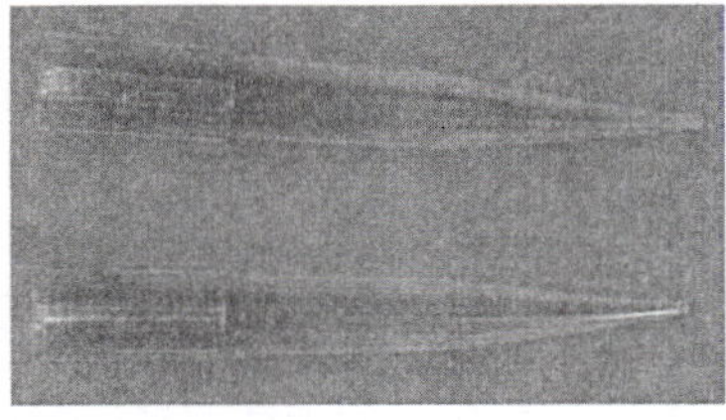

Fig. 6.6: Plastic (Blue) tips used with single channel pipette.

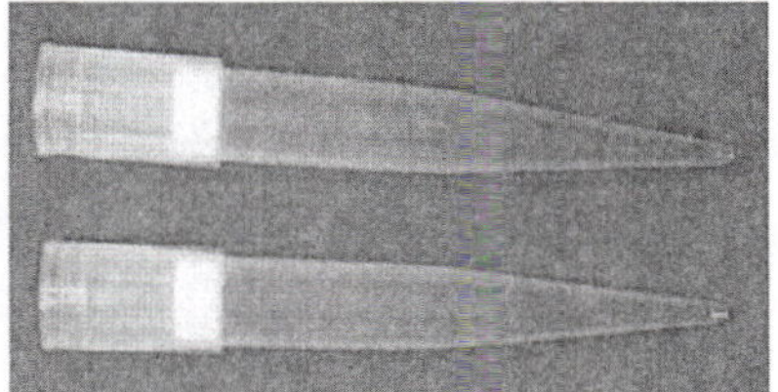

Fig. 6.7: Aerosol plastic tips.

(a) *Fixed volume pipette:* A fixed volume pipette is used to dispense a fixed (same) quantity/volume of liquid every time. The measuring/dispensing volume of the pipette cannot be changed.

(b) *Variable volume pipette:* A variable volume pipette is ideal for dispensing liquid in different/ changing volume whenevr needed (Figure 6.8).

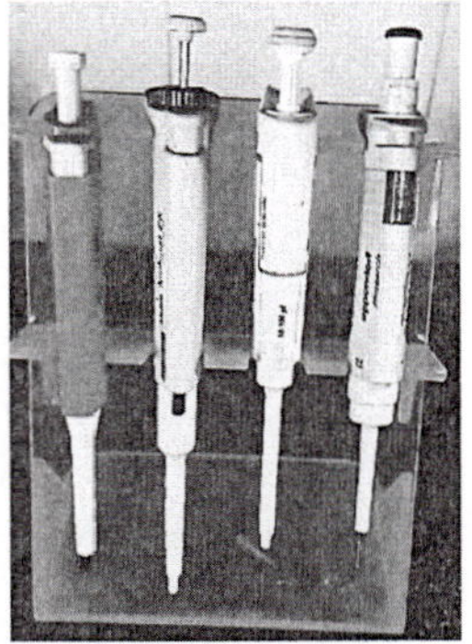

Fig. 6.8: Variable volume pipettes of different sizes.

(ii) *Multichannel Pipette:* Multichannel pipettes are commonly used for dispensing liquid components in microplate applications, such as during ELISA, PCR, or cell culture experiments. Manual multichannel pipette is generally used for medium scale experimentations. Multichannel pipettes (Figure 6.9) are available as 8- or 12-channels to work with 96-well microplates. Moreover for working with 384-well microplates, a 16-channel version is used.

However, for high-throughput experiments where a lot of repetitive pipetting is performed, electronic multichannel pipette is used for a significant ergonomic benefit.

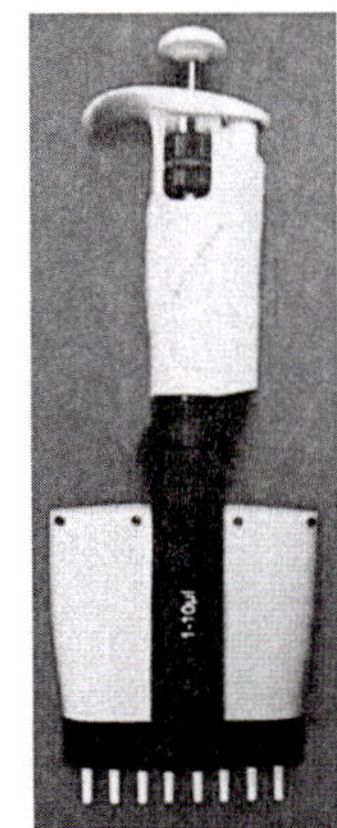

Fig. 6.9: A multichannel pipette.

Positive Displacement Pipettes

A positive displacement pipette operates via piston-driven displacement. Unlike an air displacement pipette (which dispenses liquid using an air cushion), the piston in a positive displacement pipette creates the aspiration force, which remains constant (Figure 6.10).

Handling of Pipette

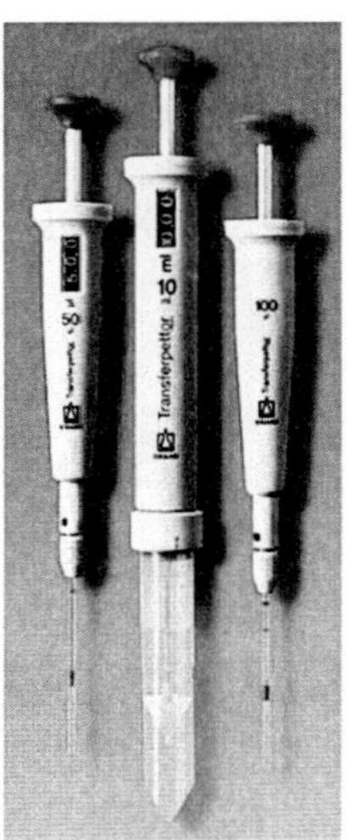

Fig. 6.10: Positive Displacement Pipettes.

1. Set the volume to be dispensed using the pipette,
2. Fit an appropriate dispensable tip tightly on the pipette,
3. Push the knob/piston down until the first-step only,
4. Immerse the tip into the liquid to a proper depth before aspiration,
5. Release the plunger slowly at constant speed,
6. Pull the pipette straight out,
7. Examine the tip (containing liquid) before dispensing out the sample,
8. Insert/place the tip of the pipette into the tube's wall and press the plunger till the second step,
9. Examine the tip (became empty) after dispensing the liquid sample,
10. Discard the tip after dispensing the liquid and use a fresh/autoclaved tip to avoid contamination.

Precautions

1. Use an appropriate pipette, according to the volume to be dispensed,
2. Use the correct tip, according to the pipette,
3. Ensure that the pipette tip is tightly fitted on the pipette,
4. Eject the tip by pressing tip-ejection knob,
5. The pipette is ready for next round of use.

Uses of Pipettes

- To dispense a small amount of solvent or solution.
- For serial dilutions, measuring/dispensing liquid.
- For biological experiments and laboratory tests for measuring and dispensing media, reagents and cell culture.
- Measuring/dispensing buffers and enzymes.
- Preparing PCR mix.

- Measuring and dispensing organic solvents.
- Aliquoting DNA, bacterial culture, plasmid DNA, etc. in biotechnological experiments.
- Dispensing smaller amount while adjusting preparing buffer and pH adjustment.

Let's Check Your Understanding

1. Pipette is fitted into vaccupet with light pressure only- they must not be forced onto the pipette. True/False
2. Who gave the earlier design of the Liebig Condenser?
   a. Justus von Liebig
   b. Weigel and Gottling
   c. Fritz Walter Paul Friedrichs
   d. Otto Dimroth.
3. In which design of vaccupets there is no requirement of squeezing?
   a. Plastic wheel design pump
   b. Wheel style pipette filler
   c. Flip style pipette filler
   d. Both a and b
4. You can release air from the pipette filler by squeezing which valve?
5. Adjusted volume in micropipette is shown in which part?
6. What are three positions of micropipettes?
7. What is the primary function of a vacupette in laboratory use?
   a. To heat liquids
   b. To create a vacuum for filling and releasing reagents from a pipette
   c. To measure the pH of solutions
   d. To sterilize laboratory equipment
8. Which component of the micropipette is crucial for adjusting the volume of liquid to be dispensed?
   a. Eject button
   b. Volume window
   c. Plunger
   d. Pipette tip
9. Which of the following is NOT a type of pipette mentioned in the chapter?
   a. Air displacement pipette
   b. Positive displacement pipette
   c. Heat displacement pipette
   d. Variable volume pipette
10. What precaution should be taken when using a micropipette to ensure its longevity?
    a. Store the micropipette with the plunger fully depressed
    b. Keep the micropipette at its maximum volume setting when not in use
    c. Store the micropipette with the volume set slightly lower than its maximum capacity
    d. Use the micropipette to dispense hot liquids occasionally

Answers

1. True
2. b
3. a
4. A valve
5. Volume window
6. Rest position, First stop, and Second stop
7. b
8. b
9. c
10. c

7

Use and Handling of Microscopes

Introduction

To accomplish experimentations in laboratories, sometimes we need microscope to test/validate our ideas/findings. A microscope is needed to examine objects which are very small which cannot be visualised by naked eyes. However, microscopic study requires some expertise in operating the microscope, particularly the sophisticated microscopes, as well as to prepare (peeling, sections, cutting, fixation, staining, dehydration, mounting, etc.) the sample, so that it can be appropriately examined for the anticipated details.

According to the scientific requirements different types of microscopes are used in the laboratories for research purpose. In this chapter, our aim is to acquaint you with the basics of using and handling of microscopes, particularly a compound microscope that is commonly used by the graduate and post-graduate students for basic research.

Microscopy is often designated as the scientific investigation that is employed to review and analyse minute structures and objects using a microscope. The term "microscope" has been derived from ancient Greek words "micros" meaning "small" and "skopeîn" meaning "to look at or see,".

Microscope

A microscope is a laboratory instrument used to examine objects (biological specimens) which are too small to be seen by the naked eye. The microscope must accomplish three tasks:

- **Magnification**- to provide a magnified image of the specimen.
- **Resolution**- to separate the details in the image.
- **Contrast**- to render the details clearly visible to the eyes.

Although we will tnroduce to you the different types of microscopes used for different speciments, our main focus is to discuss about compound microscope, which is commonly used in the laboratories. Compound microscope was developed around the beginning of 17th century.

Types of Microscopes

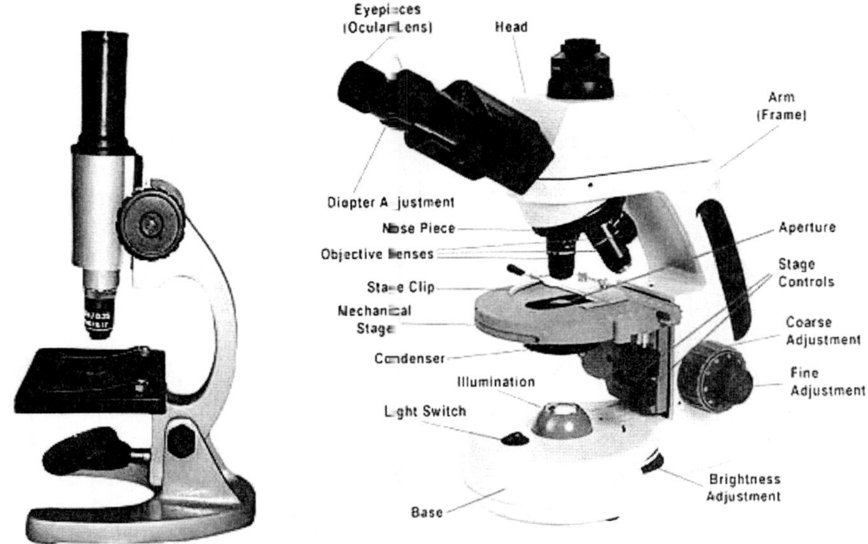

Simple microscope

Compound microscope

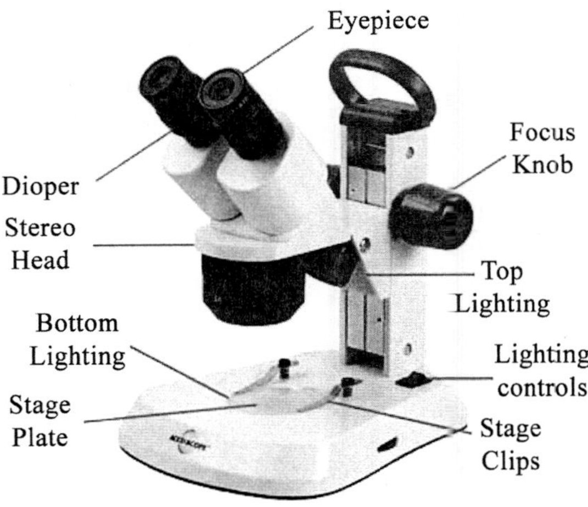

Stereo microscope

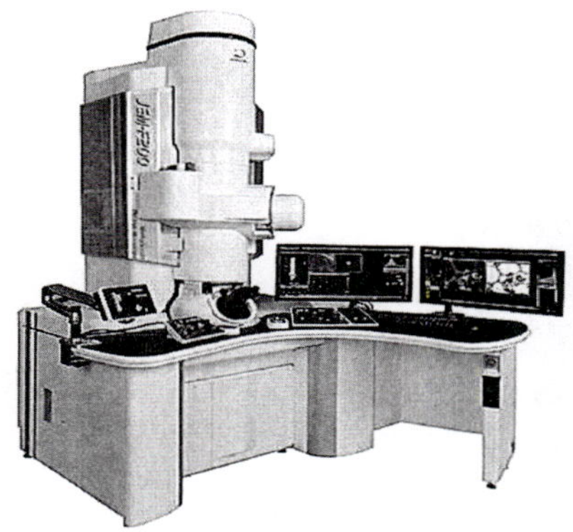

Transmission electron microscope

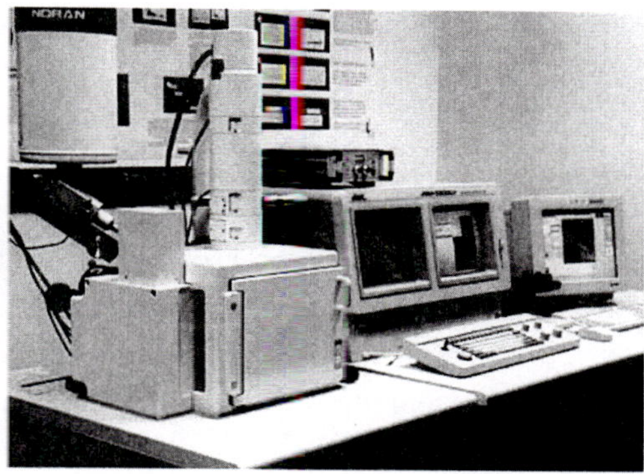

Scanning electron microscope

Fig. 7.1: Different types of microscopes.

Compound microscope consists of two convex lenses aligned in series: an objective, closer to the object or specimen, and an eyepiece (ocular) closer to the observer's eye. It achieves magnification in two-stages. The objective lens projects a magnified image into the body tube of the microscope and the eyepiece further magnifies the image projected by the objective lens. Compound microscope has higher magnification but low resolution. It is the most frequently used microscope now-a-days. The essential specifications of a compound microscope include:

- **Magnification** refers to making the specimen look larger through the microscope due to zooming in by the lenses, commonly ranges from 40× through 400× up to 1000×.
- **Resolution** refers to how well the image is captured by the lenses of compound microscope. A higher resolution provides more clear and detailed image.
- **Contrast** refers to the darkness of the background relative to the focus on the specimen. Magnificent contrast is achieved by staining the specimens.

Parts of a Compound Microscope

- **Body Tube** is the upper part of the microscope that connects the eye-piece to the objective lenses.
- **Arm** is the structural component that connects the body tube of the microscope to the base.

- **Base** is the supporting structure of microscope, which consists of electrical components for operating/controlling intensity of light to illuminate the sample.

Optical Components of a Compound Microscope

- **Objective:** Its main function is to gather the light passing through the specimen/sample and to project an accurate, real, and inverted image in the body of the microscope. A compound microscope may have 3–5 objective lenses having the magnification power ranging from 4× to 100×, which can be used one-by-one.
- **Eyepiece or ocular lens:** Its basic function is to 'look at' the focused, magnified real image projected by the objective piece. The standard magnifying power of eyepiece is 10×.
- **Condenser:** It functions to gather the light coming from the light source and concentrate/condense the light in a collection of parallel beams onto the specimen. It is present underneath the stage.

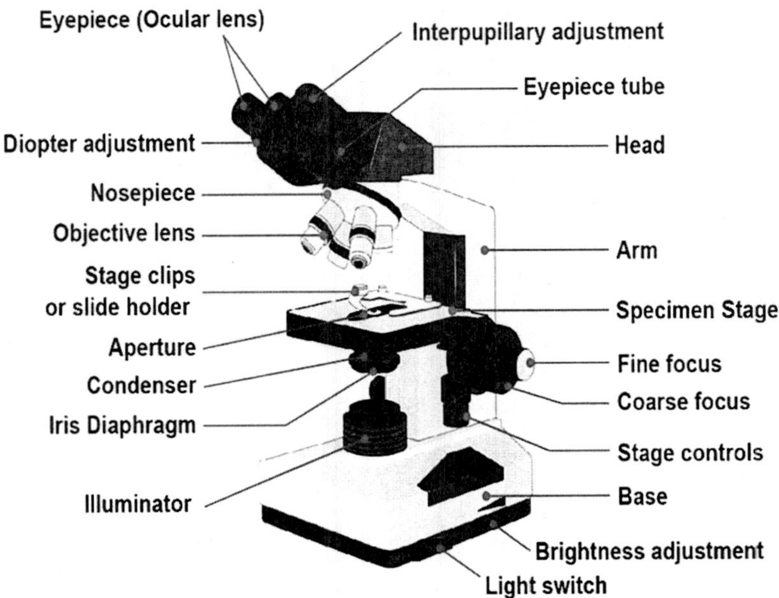

Fig. 7.2: Different parts of a typical compound microscope.

Mechanical Components

- **Diopter adjustment:** It is used to adjust the eye-piece (ocular lens) to adapt for the difference in distance between two eyes which may vary from person to person.
- **Nosepiece:** It is a rotating structure that holds the objective lenses. By rotating the nose-piece one can select an objective lens of appropriate magnification (from 4×, 10×, 40× or 100×).
- **Coarse and Fine adjustment knobs:** These knobs bring the specimen into focus by raising or lowering the stage. They are mounted on the same axis with fine focus knobs present inside the coarse focus knobs. Coarse adjustment is used for finding the focus by moving the stage rapidly, whereas fine adjustment moves the stage slowly to refine adjustment in focusing.
- **Stage:** This is the platform where the specimen slide (to be viewed under the microscope) is placed. It usually has a hole in the center that allows light to pass through the specimen. The stage can be moved (forward/backward and left/right) using the stage control knobs.
- **Stage clips/brackets:** These are used to hold the specimen slide in a position on the stage.
- **Aperture:** It is the hole in the stage through which the light from illuminator reaches the stage after being switch on/off the illuminator. condensed by the condenser.
- **Illuminator:** It is the source of light for a microscope, usually situated at the base of the microscope.
- **Iris diaphragm:** It is located between the condenser and the stage. It functions to control the amount of light reaching the specimen.
- **Condenser focus knob:** It adjusts the condenser up or down to control the light focusing on the specimen.
- **Brightness adjustment:** It is used to adjust the light brightness coming from the illuminator.
- **Light switch:** It can be present on the side or back of the base of the microscope, which is used to switch on/off the illuminator/source of light.

Working of a Microscope

Compound microscope can be used to view a magnified image of a specimen:

- The specimen is placed between the objective and condenser, after preparing it on a slide.
- Light from the light-source is pointed over the specimen with the help of a condenser.
- The light passes through the specimen and comes to the objective lenses.
- The objective piece captures the light coming from the specimen and creates a magnified image of the specimen, which is the primary image.
- The objective piece passes this image through the body tube to the ocular lens/eyepiece, which further magnifies the image.
- Finally, the viewer can see a clear and magnified image of the specimen through the eyepiece.

Focusing the microscope

To focus the microscope on a specimen, start by turning the revolving nose-piece (turret) so that the objective piece of lowest magnification/power is in the position. The lowest power objective piece is easier to focus and centre the image in the field of view. Then turn the coarse focus knob to move the stage up towards the objective piece. Care should be taken so that objective piece does not touch the slide.

Next, look through the eyepiece and adjust the illuminator and diaphragm until maximum, comfortable level of light is attained. Use the coarse adjustment and the fine adjustment knobs to bring the specimen in perfect/clear focus. Move the slide (using stage control knobs) until the image comes in the centre of the field of view. Once a clear image is obtained, one can shift to a higher power objective piece, which requires a minimum focusing adjustment. If the microscope is not having a rack stop, care must be taken not to allow the objective piece to touch the slide, as it may cause breaking of slide. Remember that, the microscope forms an inverted (upside down/back-to-front) image of the specimen. When we move the slide to right, the image moves to the left and vice-versa.

Magnification of the specimen obtained using a microscope can be calculated using the power of objective and eyepiece, as follows:

Magnification= Power of objective × Power of eyepiece

Use / Handling of a Microscope

- **Handle with care:** While carrying a microscope from one place to another, hold it by the base and the metallic arm to avoided misalignment. Use both the hands to carry a microscope. While transporting it, one must use a microscope bag/box provided with it. Always keep the microscope in an upright position. Care should be taken to avoid any kind of bump on it.
- **Keep objective lenses away from slide:** While adjusting the focus, let the objective piece go down as far as it can, but avoid touching the slide containing specimen. Do not use distilled water or any organic solvents for cleaning objective lens, as these may dissolve the coating on the lens. Excessive cleaning might also be harmful to the lens.
- **Clean after using immersion oil:** If immersion oil is used while visualizing the image, ensure that the objective pieces are cleaned immediately after the use by using lens paper and lens cleaner, not the solvents. Do not touch any lens with bare hands, because the oil and dirt on fingers can abrade the glass/lens. Paper towels/toilet papers have fibers which can stick to the lens and they can scratch the lens. Therefore, microscope lens should be cleaned using lens paper only.
- **Cover when not in use:** The microscopes must be covered with cover provided with the microscope, when not in use, to protect it from dust.
- **Care of the bulb:** After using the microscope, illuminator bulb must be turned off as soon as microscopy is complete to let the bulb cool down before it is covered and to save the life of the bulb.
- **Store in a clean, dry place:** Avoid keeping/storing microscopes in an area having corrosive chemical fume or poor ventilation, as these can destroy lens or metal parts of the microscope.
- **Cleaning of microscope:** Use an aspirator to remove dust while cleaning after storage of microscope. Use appropriate paper/wipes for cleaning the lenses to avoid scratches on lens.
- **Keep the User's Manual and wrenches in a near/safe place:** Consult with the User's Manual before making any adjustments in the microscope.
- **Conduct an annual maintenance check:** One should perform annual maintenance checks of the moving parts of the microscope, once in a year, to keep it clean and lubricated.

- **Professional servicing of microscope:** For every 200 hours of use or at least once in 3 years, whichever comes first, the microscope should be got serviced by the professional/manufacturer of the microscope.
- **Adjustment of eyepiece:** Adjust the eyepiece in such a way that it angles toward the user, by doing so one gets the adjustment knobs on right-hand side. Do not use force on any adjustment knobs.
- **Focus carefully/properly:** While focusing on the specimen, do not be fast/harsh and focus gently on the specimen. Take care that objective piece does not touch the slide.

Handling after use of the microscope: After finishing the work/microscopy, take out the slide from the stage, keep the lowest power objective towards the stage. Keep the illumination bulb turned off when the microscope is not in use, as bulb is very expensive and it has a definite working life. Disconnect electric power supply and cover the microscope with dust cover provided with it.

Let's Check Your Understanding

1. The ability of microscopes to produce a clear and detailed image is called its ________________.

   a. Magnification   b. Resolution

   c. Contrast   d. Focal Length

2. The total magnification for each objective lens is calculated by the following formula.

   a. Magnification= Power of Eyepiece × Objective

   b. Magnification= Power of Eyepiece / Objective

   c. Magnification= Power of Objective / Eyepiece

   d. Magnification= Focal Length × Distance

3. The condenser regulates the amount of light passing from the light source through the specimen and the lens system of the microscope and provides better contrast.

   True / False

4. What can be the maximum magnification of a compound microscope?

   a. 800×   b. 500×

   c. 1000×   d. 1200×

5. Where is the diopter located on a microscope?

   a. Base   b. Eye-piece

   c. Between condenser and stage   d. Ocular tube

6. Name the two types of lenses used in a Compound Microscope.

7. Which is NOT the function of a microscope?

   a. Contrast

   b. To form a real, erect image of the specimen

   c. Resolution

   d. Magnification

8. What are the precautions required while focusing the microscope?

9. Specifications of a compound microscope is defined in terms of?

10. A compound microscope can have how many objective piece?

Answers

1. b
2. a
3. False
4. c
5. d
6. Eyepiece and Objective lens
7. b
8. Following precautions are necessary while focusing a microscope:
   - Use the lowest power objective first, and then move to higher power objectives.
   - Care should be taken to avoid touching the objective lens with slide.
   - Adjust the image formation in the center of field of view with the help of coarse/fine adjustment knobs.
9. Magnification, resolution and contrast
10. 3–5

8

Weighing of Chemicals and Preparation of Solutions

Introduction

In a research laboratory, we frequently need weigh chemicals and prepare different solutions of varying strength, as per the requirements. Accuracy in weighing and preparation of solution, with respect to weight/volume and strength, determines the success of experiment In turn, the accuracy depends on weighing of solid chemical and/or measuring the volume of liquid chemical/making up the volume. In this Chapter, weighing of chemicals using appropriate weighing balance, preparation of solutions of different strength, their dilution, and precautions while preparing will be discuss in brief.

Weighing of Chemical

The primary method to determine the amount of chemicals to be used is weighing the chemical by using a balance.

In general, there are two types of balances to weigh chemicals in a laboratory:

1. Standard laboratory balance
2. Analytical balance

Standard Laboratory Balance

Generally, a standard laboratory balance is used for weighing a chemical when an error of ±0.02 g is acceptable. It is also used for pre-weighing, where approximate mass needs to be determined. Precise mass of the chemical is then determined using an analytical balance. Following are the steps in using a weighing balance in a laboratory.

1. Clean the balance, if necessary. Turn on the balance and wait until it displays 0.0 g or Tare it to make 0.0 (which also confirms proper functioning of the balance.

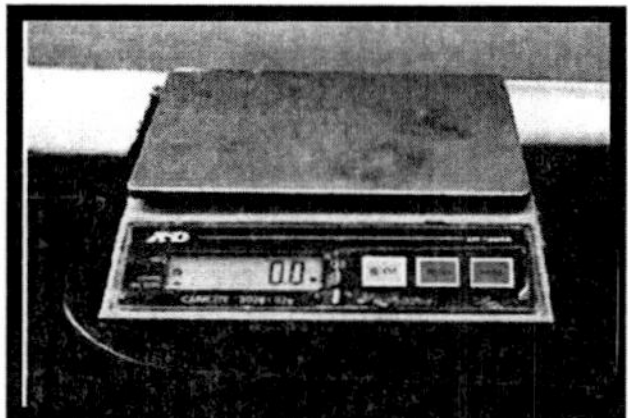

2. Place a weigh boat on the weighing platform of the balance.

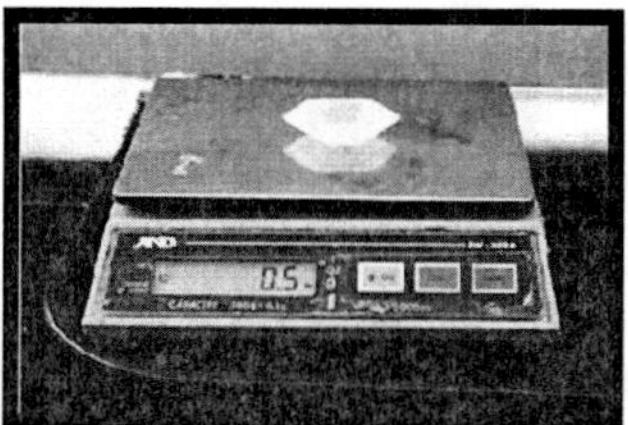

3. Press the TARE or ZERO button to bring the display at 0.0 g.

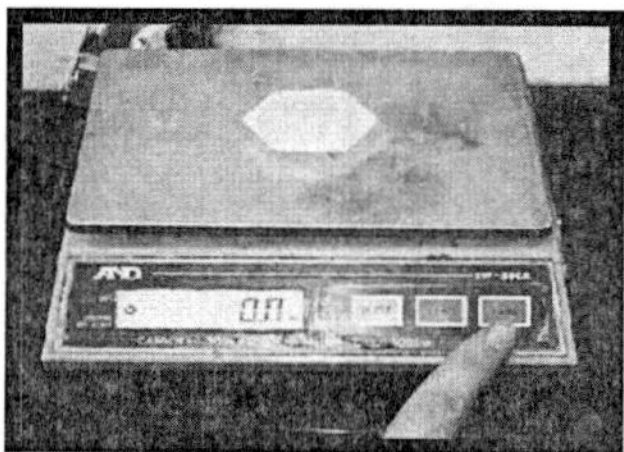

4. Place the chemical, to be weighed, in the weighing boat until the desired quantity/value appears on the display of the balance.

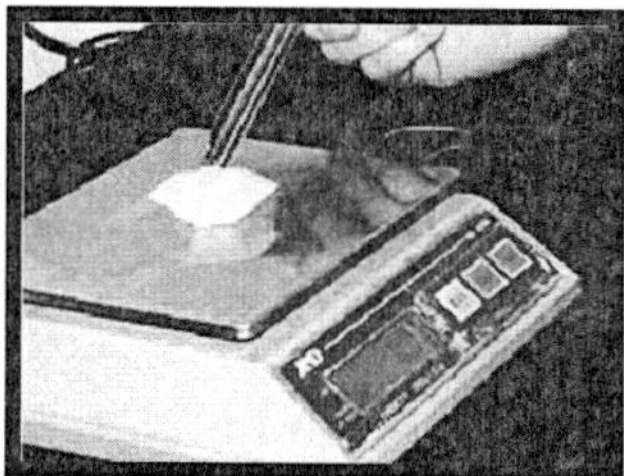

5. Record the reading and then remove the weigh boat from the weighing balance, clean the area properly and switch off the balance.

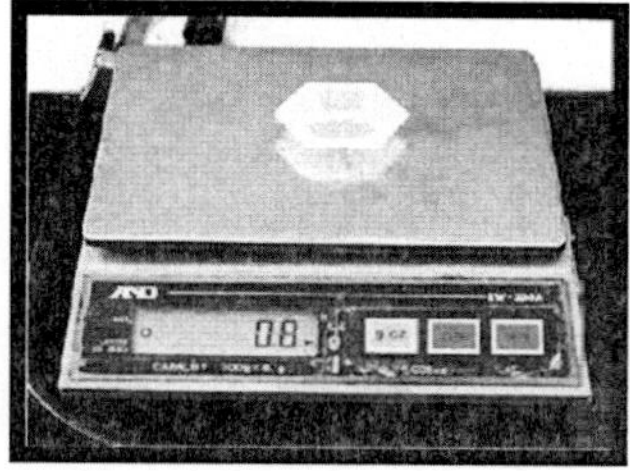

Analytical Balance

An analytical balance is very sensitive and used when weighing of a chemical requires error smaller than 0.01 g. Generally, such a balance provides the mass of chemical closer to 0.0001 g (or 0.00001 g). The balance is enclosed in a chamber/hood to protect the weighing process from external factors like air currents, etc. Before using an analytical balance, the maximum weighing limit must be noticed to avoid crossing the limit. The balance must be frequently calibrated using a specified standard check weight.

1. Clean the balance, if needed. Close all the sliding doors of the balance to avoid any air-disturbance to the balance. Turn on the balance and wait until it displays 0.0 g.

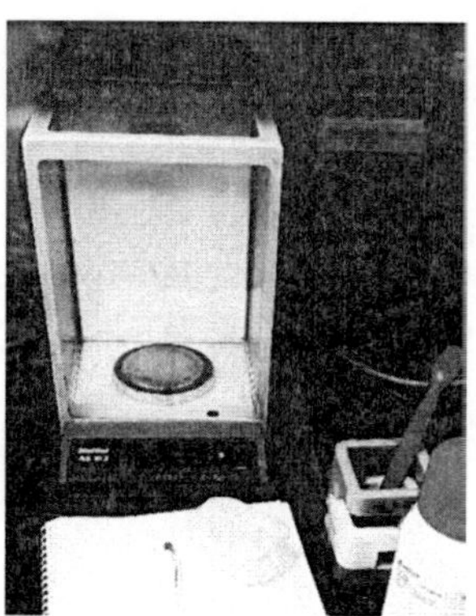

2. Open any one (or two) sliding door to place a weighing boat on the platform of the balance. Press ZERO / TARE button, wait till it comes to 0.0000. It is important to set the balance to 0.0000 otherwise it will display the combined weight of the boat and chemical.

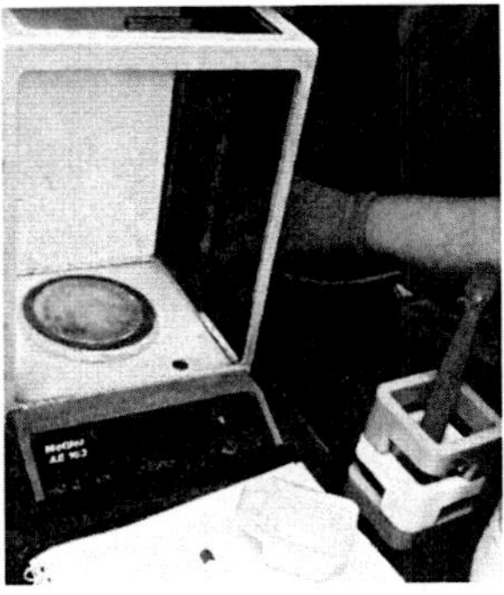

3. Place the chemical to be weighed in the centre of the weigh boat and close the chamber's door. As soon as the reading on display becomes stable, note down the reading.

4. Open the door to remove weighing boat, along with the chemical, and use the chemical for desired purpose. Clean the area inside the balance with hair brush and close all the doors of hood. Turn off the balance.

Note: Use/wear gloves and apron while weighing/handling hazardous chemicals.

Preparation of Solution

While a solution is defined as a homogeneous mixture of one or more solutes dissolved in a solvent, it is required in different strength/concentration depending on the experimental requirements. A chemical/solute is dissolved in a liquid solvent to make the solution. A wide range of solutions are used in a laboratory in different strength/concentration. Strength/concentration of a solution is the quantity of solute dissolved in the solution. Strength/concentration of a solution can be expressed in many ways like moles per litre, grams per litre, in the form of percentage, parts per million, etc.

Things Need to be Known before Making a Solution

- Desired concentration
- Desired volume

Ways to Express Strength/Concentration of a Solution

1. Molarity (m) moles per litre of solution

Molarity is one of the commonly used methods of expressing concentration of a solution. It indicates the number of moles of solute dissolved in one litre of solution, symbolised as M and the unit is moles per litre (mol/L).

Number of moles (n)= [Mass of solute in gram] ÷ [Molecular weight in gram (M)]

Molar (M) concentration of solution (mol/L)= n ÷ V

where

n= number of moles of solute

V= volume of solution in liter.

Number of moles (n) of solute in a certain volume (V) of a solution of known concentration (M) can be calculated using the formula: n= M × V

Preparation of 1.0 Molar, 1 litre Aqueous Solution

1. Calculate the amount of the chemical required to prepare 1 litre solution of given molarity (e.g., 1.0 M).
2. Weigh the amount of chemical required using an electronic balance.

3. Transfer the weighed chemical into a beaker containing about two-third of (the final volume of solution to be prepared) distilled water (~700 mL).
4. Let the solute dissolve completely in the taken volume of water (~700 mL).
5. To dissolve the chemical, a stirring rod or magnetic stirrer can be used to dissolve properly. Sometimes it may also be necessary to gently heat the mixture to speed up/complete dissolution of the chemical/salt.
6. Transfer the solution to 1 L measuring cylinder or volumetric flask, rinse the beaker, and pour into the measuring cylinder.
7. Make up the final volume (1000 mL) using distilled water until bottom of the lower meniscus comes in line with of the 1000 mL mark on the measuring cylinder.
8. Mix the content thoroughly and transfer the prepared solution to a reagent bottle.
9. Label the bottle with name of the solution, strength/concentration, and date of preparation.

2. Molality (m) moles per kilogram of solution

Molality refers to the number of moles per 1000 g (i.e., 1 kg) of a solvent. It is dependent on the density of the solvent. It differs from molarity, as M refers to the volume of the solution (which is temperature dependent). Molal solutions are generally not used nowadays in biochemical experiments.

3. Concentration by Percentage

Usually, it is expressed as % (w/v, v/v, or w/w)

Weight / Volume Percentage

When the solute is solid, solution is prepared as percentage weight per volume (% w/v). It is the mass of solute (in grams) dissolved in 100 mL of (aqueous) solution.

$$\%\frac{w}{v}=\frac{\text{Mass of solute}\,(\text{g})}{\text{Volume of solution}\,(\text{ml})}\times 100$$

Volume / Volume Percentage

When the solute is liquid, solution is prepared as percentage volume per volume (% v/v). It is the volume of solute (liquid, in ml) mixed in per 100 mL of solution.

$$\%\frac{v}{v}=\frac{\text{Volume of solute}}{\text{Volume of solution}}\times 100$$

Weight / Weight Percentage

This is the mass (g) of solute per 100 g of solution. It is commonly used in the aqueous commercial preparations like concentrated solutions of acids. Here the solution can be prepared independently of temperature considerations.

$$\%\frac{w}{w}=\frac{\text{Mass of solute}}{\text{Mass of solution}}\times 100$$

4. Concentration in Grams per Litre

A solution can also be prepared by dissolving a solute of known mass (g) in a given volume of solvent so that the concentration is expressed as grams of solute dissolved in one litre of solution.

5. Concentration by Parts per Million (ppm)

Parts per million (ppm) indicates how many parts of a certain molecule/compound is present in one million parts (10,00,000, Ten lakhs) of the solution. It is generally used to express the concentration of chemicals dissolved in a solvent (usually water) or a compound in soil.

6. Saturated Solution

A solution is called saturated if no more solute can be dissolved in the solvent at a given temperature (generally room temperature 25 °C). Solubility of solid solute depends on the type of solute, type of solvent, and temperature. Solubility of a chemical compound can be increased by increasing the temperature. Before making a saturated solution of a chemical, it is important to know its solubility. This can be obtained from a solubility curve, which differs for different chemicals.

Dilution of Solution

Dilution is the process in which concentration of the solute in a solution is decreased by adding more solvent.

Three things that occur on dilution:

1. Solvent is added
2. Concentration decreases
3. Volume increases

Dilution related calculations can be made using the formula:

$S_1V_1 = S_2V_2$

where: S_1= strength/concentration of original solution, V_1= volume of original solution, S_2= strength of diluted solution,V_2= volume of diluted solution.

Types of Dilution

A. Simple Dilution

In this dilution, one unit volume of solution is added with an appropriate volume of solvent to get the desired concentration/final volume. For example, a 1:5 dilution can be made by taking 1 unit volume of the solution added with 4 units of diluent (solvent) to make the final volume 5 (dilution factor).

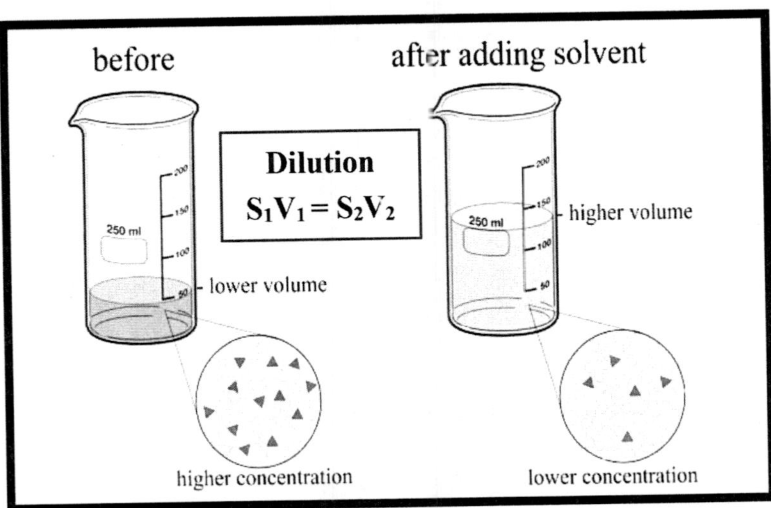

Fig. 8.1: Simple dilution.

B. Serial Dilution

Serial dilution is a stepwise dilution of a substrate in solution, which amplifies the dilution factor wherein small initial quantity of solution is taken and mixed with appropriate volume of diluent.

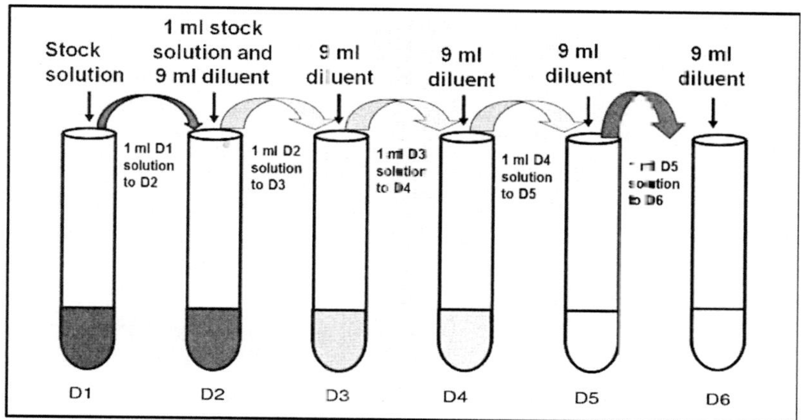

Fig. 8.2: Serial dilution

Dilution factor of a tube in a series

$$Dilution\ factor = \frac{\text{Vol. of sample taken}}{\text{Vol. of sample} + \text{Vol. of diluent}}$$

For Example

Dilution factor for the first tube = 10^{-1} (1 mL solution added to 9 mL of diluent)

Dilution factor for the second tube = 10^{-1} (1 mL solution added to 9 mL of diluent)

Dilution factor for the last tube = Dilution factor for the previous tube × dilution factor for the last tube = $10^{-4} \times 10^{-1} = \mathbf{10^{-5}}$

C. Making a fixed volume of specific concentration from liquid reagents

$V_1C_1=V_2C_2$ or $(C_1) \times (V_1)=(C_2) \times (V_2)$

where

C_1 and V_1 are source solution attributes

C_2 and V_2 are new solution attributes

C = concentration

V = volume.

For Example

To prepare 50 mLl of 0.1 M NaOH solution, how much 1.0 M stock solution of NaOH will be required ?

First of all find out:

C_1 = 1.0 M

V_1 = ?

C_2 = 0.1 M

V_2 = 50 mL

Formula to be used

$(C_1)(V_1) = (C_2)(V_2)$

$(1\ M) \times (V_1) = (0.1\ M) \times (50mL)$

$V1 = [(0.1M) \times (50\ mL)] \div (1M)$

= 5 mL (**Answer**)

Let's Check Your Understanding

1. Weighing is a method used to determine ……. …….. of a substance/ chemical.
2. Solution is a heterogeneous mixture of one or more than one solute dissolved in a solvent.

   True / False
3. The concentration of the solute ……… and volume of solution ……. upon dilution.
4. What are the two different types of balance used?
5. List different type of dilution.
6. When an error <0.01 g is required then which weighing balance should be used?
7. What is molarity?
8. What are the two major parameters need to be determined before making a solution?
9. True or False: An analytical balance is enclosed in a chamber to protect the weighing process from environmental factors such as air currents.
10. True or False: The concentration of a solution expressed in parts per million (ppm) indicates the number of parts of a solute per one hundred thousand parts of the solution.

Answers

1. Mass
2. False
3. Decreases and Increases
4. Standard laboratory balance and Analytical balance
5. i. Simple Dilution,

   ii. Serial dilution.
6. Analytical balance
7. Number of moles of solute dissolved in a litre of solution.
8. The desired concentration and desired volume.
9. True
10. False

9

Preparation of Different Agrochemicals

Introduction

For agricultural applications as well as for research work, often we need to prepare solutions of different agrochemicals of varying strength, as per the requirements. Agrochemicals are used to help crop plants grow, produce, and protect them from biotic/abiotic stresses. Many of the agrochemicals (insecticides, herbicides, fungicides, rodenticides, molluscicides, nematicides, etc.) are used to protect the crop from insect pests and diseases. However, some of them (fertilizers and soil conditioners) are used to augment crop yield. While in the previous chapter (Chapter 8) we discussed about preparation of laboratory solutions, in this chapter we will briefly discuss about the preparation of different agrochemicals.

Agrochemical/agrichemical is a generalized term used for the chemicals applied in agriculture. In general, it also includes the feed-supplements used in fisheries and animal husbandry.

Types of agrochemicals

Type	Description	Examples
Pesticides	Chemicals or substances used to control pests, which are harmless to cultivated plants or to animals.	Fungicides, insecticides, herbicides, etc.
Insecticides	Used to destroy insects. Insecticides can be ovicides (kill eggs), larvicides (kill larvae), etc.	Organo-chlorines, organophos-phates, carbamates, and pyrethroids.
Herbicides	Used to kill weeds/herbs.	Gramoxone and glyphosate
Fungicides	Used to control fungi and oomycetes.	Mankocide
Algaecides	Used for controlling algae.	BioGuard Algae All 60
Rodenti-cides	Used to prevent spread of rodents like rat, mice.	Klerat
Mollusci-cides	Used to control molluscs like snails and slugs.	Slugit
Nemati-cides	Used to control nematodes.	Furadan

Agrochemicals

Fertilizers

Fertilizers are used to augment growth of crop plants and to mitigate nutrient deficiency in the soil. Fertilisers can be organic (containing natural substances, prepared through natural processes) and inorganic fertilisers (synthetic fertilizers, manufactured artificially using chemical processes using natural deposits, by chemical alterations).

Soil Conditioners

Soil conditioners include manure, livestock and leaf compost, peat that help making the soil in good condition. They are laid on the top soil and then mixed properly. They not only increase nutrient contents in the soil but also improve aeration and water holding capacity.

Liming and Acidifying Agents

Sometimes soils can be too acidic or too alkaline for proper growth and development of crops. Under such circumstances, liming and/or acidifying agents are used in the soil to adjust its pH. When the soil is too acidic, calcite in the form of powdered limestone is applied. Whereas in alkaline soil, sulphur compounds are added to neutralize it.

Preparation of Doses

The most common agrochemical are the herbicides, which account for approx. 80% of the pesticides. The commercial herbicide contains the herbicide chemical (active ingredient), adjuvants as well as filler materials. The percentage of adjuvants and other inert/filler materials depends on formulation and the manufacturing company. Therefore, it is necessary to prepare agrochemicals based on active ingredient or acid equivalent.

Active Ingredient

The chemical in commercial product that is directly responsible for the herbicidal activity is called active ingredient (a.i.). Active ingredient of each herbicide formulation is displayed by the manufacturer on the container used for its packaging. It is generally expressed as percent by weight or volume.

Types of Formulations

Dust: Prepared by mixing toxicant with the dust carrier.

Finished product = Toxicant + Carrier

(0.1–25%) (99.9–75%)

Solution concentrate: Formulated in a liquid phase (water or in oil).

In water: Monocrotophos, Organophosphate

In oil: Flint (domestic spray)

Wettable powder (WP): It is a water-soluble formulation which yields stable suspensions when diluted with water. Water insoluble toxicant is mixed with carrier. The active ingredient in WP ranges from 15–95%. While preparing WP formulation, percentage strength of spray is made on weight of pesticide to weight of water basis. Since one litre of water weighs one kg, preparation can be made using the formulas:

$$pH = \log \frac{1}{\left[H^{+}\right]}$$

For example

Calculate the amount of Malathion (57% EC) that would be required to prepare one litre solution of 2% spray (active ingredient) to control household fleas?

Calculation

Malathion (57% EC) ml = (2 ÷ 57) ×1000

= (0.035) ×1000

= 35 mL

Fertilizers commonly used in agriculture

Lack of knowledge about the type/dose of fertilizer might be one of the reasons for insufficient/excessive use of fertilizers in crops. Recommendation for nitrogen (N), phosphorus (P), and potash (K) are made as kg of N, P_2O_5 and K_2O. Available N, P_2O_5 or K_2O content in a fertilizer need to be considered while calculating the amount of fertilizer required for application in the field either as basal or top/spray application.

For example

1. Calculate the amount of urea, Single superphosphate and Muriate of potash required to be applied in one hectare of paddy with the recommendation of 120:40:40 kg N, P_2O_5 and K_2O per hectare. Whole the Nitrogen needs to be applied in three equal splits, only 1/3 N, whole P2O5 and K2O will be applied as basal dose. Given that Urea contains 46% nitrogen, SSP contains 16% P_2O_5 and MOP contains 58% K_2O.

Calculation

Urea required for 1 hectare = (100÷46) × 120 = 261 kg

Urea for the basal application = 261÷3 = 87 kg

SSP required for 1 hectare = (100÷16) ×40 = 250 kg

MOP required for 1 hectare = (100÷58) ×40 = 69 kg

Therefore, recommendation is 87 kg Urea, 250 kg SSP and 69 kg MOP as basal application.

Rest 174 kg Urea will need to be applied in two equal splits.

2. Calculate the amount of urea, DAP and Muriate of potash required to be applied in paddy field of 10 sq. m area with the recommendation of 120:40:40 kg NPK per hectare. Nitrogen needs to be applied in 3 equal splits. DAP contains 18% nitrogen and 46% P_2O_5. 1 ha= 10,000 sq. m.

Calculation

Area of field/plot= 100 sq. m= 0.001 ha

Amount of DAP required to supply 40 kg P_2O_5 will be calculated first, followed by calculation of the amount of nitrogen supplied by DAP. Then, the rest of N to be supplied by urea will be calculated.

DAP required to supply P_2O_5 (40) in 0.001 ha plot= (100÷46) ×40 ×0.001 = 0.087 kg = 87 g

N supplied by 87 g DAP = (18÷100) ×87

= 15.66 g

N to be supplied by Urea = 120–15.66 = 104.34 g

Urea required for the plot= 100÷46 ×24.34 = 52.9 g

Urea required for the plot= 100÷46 ×80 = 173.9 g

MOP required for the plot = 100÷58 ×40 ×0.001

= 0.069 = 69 g

Therefore, 87 g DAP, 53 g Urea and 69 g MOP as basal dose and rest 174 g Urea in two equal splits.

Pot Application of Fertilizers

Two ways can be applied to calculate fertilizers for pot applications. Either the surface of pots can be calculated or the weight of soil in one ha land and that in the pots. The weight of soil in 1 ha depends on the bulk density of the soil, the nature of soil (mineral or organic, its organic matter content, etc.). For a typical mineral soil surface of one hectare–furrow slice weigh ~2 million kg.

For example, if we want to determine the need of NPK at 120-60-40 kg/ha for pot application:

120 kg N in 2000000 kg of soil

120000 g N in 2000000 kg

or 120 g in 2000 kg of soil

120000 mg N in 2000 kg of soil

or 120 mg N in 2 kg of soil

or 60 mg N in 1 kg of soil

Thus, we can also calculate the requirement of P and K fertilizers. Depending on the source of N like Urea, P_2O_5 (DAP or SSP), etc., we can calculate the required amount of fertilizers for pot experimentation/application.

Disadvantages of using Agrochemicals

- Improper/excessive usage of agrochemicals may cause different types of pollution in the environment. Moreover, spray drift in the surrounding may contaminate the user, public, and/or the produce as agrochemical residues.

- Bioaccumulation and biomagnification of agrochemicals, generally not found in the living body, may get accumulated or further increased in concentration in the body, which might not be removed from the body or cannot be broken-down/excreted from the body; hence, get stored in the body/tissues. This is harmful to the organism or even to the predator consuming the organism.

- Agrochemicals (insecticides and pesticides) disturb biodiversity by affecting various organisms like microbes, friendly insects, pollinators, fish, etc.
- Poisonous agrochemical residues in food grains might be consumed by the people unknowingly, which may lead to death/disorders.
- Accidental exposure to agrochemicals may lead to skin problems, cancer or other symptoms.

Suggestions for the Safe use of Chemicals Include

- Appropriate training to the user of agrochemicals should be ensured for both the chemical and the equipment.
- To avoid the risk of spills/splashes, use of decanting kit should be ensured, particularly while preparing/ mixing/using hazardous agrochemicals.
- Prepare only the required quantity of agrochemical formulation for use at the time.
- The preparation, decanting, and mixing should be performed in well-ventilated premises. Appropriate personal protective equipment (PPE, e.g., chemical-resistant gloves, overalls, goggles, etc.) should be used while preparing the agrochemicals.
- Follow the instructions mentioned by the manufacturer of the agrochemical on the label/packaging.
- Wear appropriate P2 facemasks or a P3 respirator for volatile agrochemicals. The respirator cartridges must provide multi-level gas protection.
- Avoid using left-over agrochemicals on non-target plants or pests.
- Equipment must be cleaned/rinses 2–3 times after chemical application. The cleaning/rinsing water must be disposed appropriately/responsibly.

Let's Check Your Understanding

1. The chemical in a commercial formulation that is directly responsible for pesticidal activity is known as ....
2. Recommendation of a agrochemical for use is made based on .... present in that formulation.
3. The chemical used in agricultural practices to aid plant/crop growth and safety is called ....
4. Percentage nitrogen found in Urea is .... and in DAP is ....
5. What can be three different types of formulations?
6. How many grams of Malathion (40% WPP) would be required per liter of water to make a 2% spray (on active ingredient basis) for controlling household fleas?
7. How much Perfekthion (20% EC) would be needed per hectare, if the recommendation for mites is 0.2 kg a.i. Perfekthion per hectare?
8. PPE needs to be used while using agrochemicals.

   True / False

9. The left-over pesticide preparation should be sprayed in others field instead of disposing it off in the drainage.

   True / False

10. While spraying a agrochemical on the crop, there is no need of using respiratory mask and/or goggles.

    True / False

Answers

1. Active ingredient (a.i.)
2. Active ingredient or acid equivalent
3. Agrochemical
4. 46% and 18%
5. Solid, liquid and gas
6. Grams of Malathion 40% WP needed per litre of water= 2%×1000/40%= 20/0.4= 50 g
7. Litres of Perfekthion 20% EC needed= 0.2/0.2= 1 litre
8. True
9. False
10. False

10

Measurement of pH and Preparation of Buffer

Introduction

Measurement of pH of a solution and preparation of different types/strength of buffers are common practices in a laboratory, particularly in the Biochemistry and Molecular Biology laboratories. Therefore, in this chapter you will be acquainted with the procedures of measuring pH of a solution as well as preparation of some of the commonly used buffers in a laboratory.

Potential or Power of Hydrogen Ion (pH)

Measurement of pH of a solution is measuring the molar concentration of hydrogen ion $[H^+]$ in the solution, which is an indication of the acidity or basicity of the solution. The symbol pH stands for "power of hydrogen" and numerical value for pH is just the negative logarithm of the molar concentration of hydrogen ions $[H^+]$. The ionic product of water (K_w) is the basis for the pH scale. The total hydrogen ion concentration from all the sources and it can be experimentally measured.

The pH scale designates the concentration of H^- (and thus of OH^-) in an aqueous solution, which ranges between 1.0 M H^+ and 1.0 M OH^-. The term pH is defined by the expression

$$pH = \log \frac{1}{\left[H^+\right]} = -\log\left[H^+\right]$$

For a neutral solution at 25 °C, in which the concentration of hydrogen ions is 1.0×10^{-7} M, the pH of the solution can be calculated as follows:

$$pH = \log \frac{1}{1.0 \times 10^{-7}} = long\,(1.0 \times 10^{7})$$

$$= \log 1.0 + \log 10^{7} = 0 + 7 = 7$$

Thus, the ionic product of water makes it possible to calculate the concentration of H^+, with given concentration of OH^-, and vice versa. The value of 7 for pH of a neutral solution is not an arbitrarily chosen figure; it is derived from the absolute value of the ionic product of water. A solution having pH greater than 7 is basic, and the concentration of OH^- is greater than that of H^+. A solution having a pH less than 7 is acidic.

Note that the pH scale is logarithmic, not arithmetic. For example, two solutions differing in pH by 1 unit, means that one solution has 10 times the H^+ concentration of the other; but it does not tell us the absolute magnitude of the difference. In other words, for every 10-fold change in concentration of the H^+ (e.g., 0.1 to 1.0), the pH changes by 1 unit.

Measurement of pH is one of the most important and frequently used procedures in biochemical laboratories. The pH affects the structure and activity of biological macromolecules; such as proteins and the catalytic activity of enzymes. Measurement of pH of blood and urine is commonly used in medical diagnoses.

Nearly everything around us is either an acidic or basic in nature, with the exception of water. Pure water is neither acidic nor basic, rather it is neutral. The pH scale was developed to measure how much acidic or basic a substance is. This scale measures pH from 0 – 14. Acids have pH values between 0 and 7, while bases have pH values between 7 and 14. Thus, pure water is neutral and has the pH value exactly 7.

The pH of an aqueous solution can be measured approximately using various indicator dyes, including litmus (or pH) paper, phenolphthalein, and phenol red, which undergo color changes as a proton dissociates from the dye molecule. For accurate determination of pH of a solution in a laboratory, a pH meter is used. The pH meter possesses a glass electrode that is selectively sensitive to H^+ concentration, but insensitive to Na^+, K^+, and other cations. In a pH meter, signals from an electrode is amplified and compared with the signal generated by a solution of accurately known pH.

The pH scale

	$[OH^-]$ concentration (mol/l)	pH	$[H^+]$ concentration (mol/l)		
1×10^{-14}	0.00000000000001	0	1	1 x 100	
1×10^{-13}	0.0000000000001	1	0.1	1×10^{-1}	Increasing acidity
1×10^{-12}	0.000000000001	2	0.01	1×10^{-2}	
1×10^{-11}	0.00000000001	3	0.001	1×10^{-3}	
1×10^{-10}	0.0000000001	4	0.0001	1×10^{-4}	
1×10^{-9}	0.000000001	5	0.00001	1×10^{-5}	
1×10^{-8}	0.00000001	6	0.000001	1×10^{-6}	
1×10^{-7}	0.0000001	7	0.0000001	1×10^{-7}	Neutral
1×10^{-6}	0.000001	8	0.00000001	1×10^{-8}	
1×10^{-5}	0.00001	9	0.000000001	1×10^{-9}	
1×10^{-4}	0.0001	10	0.0000000001	1×10^{-10}	
1×10^{-3}	0.001	11	0.00000000001	1×10^{-11}	
1×10^{-2}	0.01	12	0.000000000001	1×10^{-12}	Increasing basicity
1×10^{-1}	0.1	13	0.0000000000001	1×10^{-13}	
1 x 100	1	14	0.00000000000001	1×10^{-14}	

Fig. 10.1: The pH scale.

Measurement of pH is one of the most important and frequently used procedures in a laboratory, particularly in biochemistry. The pH affects the structure and activity of biological macromolecules; such as the catalytic activity of enzymes and shape and structure of several biological macromolecules. Measurements of the pH of blood and urine are commonly used in medical/clinical diagnostics.

The pH of an aqueous solution can be measured in a variety of ways. The most common way used now-a-days is the use of a pH-sensitive glass electrode, one of the most important part of a pH meter. However, alternative methods for measuring the pH of a solution include the use of (i) litmus or pH papers and (ii) Indicator dyes.

Litmus is an extract from specific lichen which is converted into a powder and then used in a chemical solution to treat/prepare a paper sheet, cut into strips, and thus made available in the form of litmus paper. When litmus paper is dipped into a solution to check pH, it changes colour depending on the pH of the solution. The colour of wet litmus paper is matched with colours on a colour-chart to infer the approximate pH value of the solution under test.

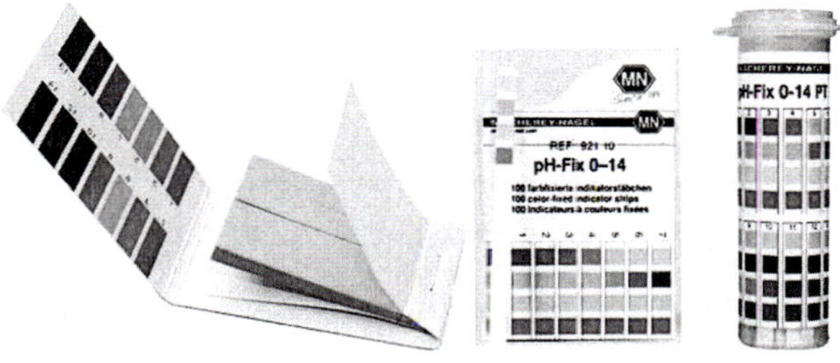

Fig. 10.2: pH papers.

The pH paper is available commercially for narrow pH ranges (e.g., 3.0 – 5.5 pH, 4.5 – 7.5 pH and 6.0 – 8.0 pH) to wider pH ranges of (e.g. 1.0 – 11.0 pH). Also, three different types of litmus papers (viz. blue, red and neutral) are available in the market. Blue litmus paper turns red in acidic solutions, and remains blue in alkaline solutions. Red litmus paper turns blue in alkaline solutions, and remains red in acidic solutions. However, neutral litmus paper turns red in acidic solutions and turns blue in alkaline solutions. At neutral pH, none of the litmus papers change colour.

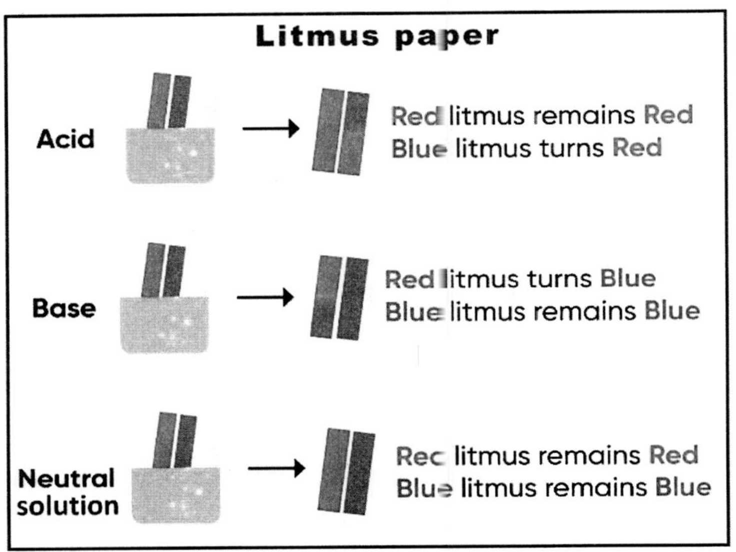

Fig. 10.3: Colour changes on litmus papers (See Colour Version on page 279)

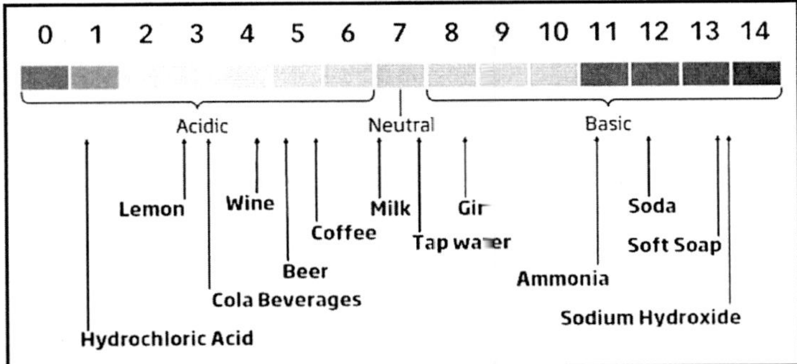

Fig. 10.4: pH of different solutions. (See Colour Version on page 279)

The litmus or pH paper cannot be used for continuous monitoring of pH change. Though pH paper is fairly inexpensive, it may be affected by certain solutions, which may interfere with the colour change. The pH indicators (also known as Universal indicators) are the materials that change colour when exposed to solution of different pH value. They are generally used in acid-base titrations. The pH indicators include Phenolphthalein, Methyl red, and Bromothymol blue, etc.

(i) Indicators: Indicators are the materials that change colour when exposed to a solution of different pH. They are used in acid-base titration, which include Phenolphthalein, Methyl red, and Bromothymol blue. Phenolphthalein turns colourless in acidic solution and becomes pink in a basic solution. It adopts four different states in aqueous solution: under strong acidic condition, it exists in protonated form, showing orange colour; under acidic condition, its lactone form is colorless. Under basic condition, it shows singly deprotonated phenolate form and gives pink colour. However, in strongly basic solution (pH >13.0), it becomes completely colourless.

Methyl red is a dark red crystalline powder and it turns red in acidic solution having pH <4.4, while it turns yellow in pH >6.2, and orange at pH between 4.4 and 6.2. Bromothymol blue is generally used for detecting neutral pH. For example, a common use of Bromothymol blue is for detecting the presence of carbonic acid, which is a weak acid making it protonated (at <pH 6.0) to turn yellow, otherwise at >7.6 pH it is deprotonated and turns blue.

(ii) pH meter: An instrument used to measure the concentration of hydrogen-ion in an aqueous solution is called pH meter. It measures the difference in electrical potential between a pH electrode and a reference electrode, and display the result converted into the corresponding pH value.

Electrodes are the key parts (rod-like structures) usually made of glass with a bulb containing sensor at the bottom of the electrode. Frequent calibration of pH meter using standard (known pH) solutions ensures the accuracy of the instrument. The pH meter might be a simple (less inexpensive, pen-like devices) to a complex and expensive laboratory instrument with computer interface and input device for temperature measurement to adjust variation in pH caused due to fluctuating temperature. Special meters and probes are available for specific applications, harsh environment, etc.

For very precise work with maximum accuracy, the pH meter should be calibrated before each measurement. For normal usage, calibration should be performed at the beginning of the day. This is because the glass electrode may not provide a reproducible measurement over longer period of time. Calibration should be performed with at least two standard buffer solutions that span the range of pH value to be measured. For general purposes, standard buffer of pH 4.00 and pH 10.00 are used.

A pH meter has a calibrate button/knob to begin calibration of the meter (setting the meter to read the pH value of the standard buffer accurately), Other control knobs are used to adjust the meter reading to the value of the standard buffer. A third control knob might also be there to allow the temperature to be set.

Standard buffer sachet/solution are commercially available from suppliers. For more precision in measurements, calibration using three buffer solutions is preferred. As pH 7 is essentially a mid-point (zeroing point), calibration should start with pH 7 first, followed by calibrating for the pH value closer to the point (pH) of interest (either 4 or 10), and finally checking for the third point provides better accuracy. After each measurement, the electrodes must be rinsed with distilled water to remove traces of the solution adhering to the wall of electrode. The electrodes should be blot dried with a clean wipe to absorb/remove water.

Measuring pH of a Solution

1. Rinse the electrode with distilled water and blot dry it with Kimwipes. Place the electrode in the test sample, press the measure button and leave the electrode in the sample for approximately 1 minute.
2. Once the reading has stabilized, press the measure button and note down the pH value of the test solution.
3. Rinse the electrode with distilled water and blot dry it with a lint-free Kimwipes. Readings of other samples can be taken or the pH meter can be kept for later use.

Measuring pH with Accuracy

To ensure correct use of pH meter, the pH meter should be regularly calibrated and maintained well for accuracy in pH measurements.

Taking / noting-down the pH reading: Be sure to wait for a while to the pH reading/value to become stable before taking the reading. Do not forget to rinse the electrode with distilled water after every measurement.

Be aware of the temperature: The standard temperature for the operation of pH meter is 25 °C. While using pH meter, ensure that the room where pH meter is used the temperature is not too high or too low.

Buffer

Buffer is an aqueous solution consisting of a mixture of a weak acid and its conjugated base, which resist any drastic change in its pH when a small amount of acid or base is added to it.

$$CH_3COOH \rightleftharpoons H^+ + CH_3COO^-$$

Acetic acid (CH3COOH), a proton donor, and the acetate anion ($CH3COO^-$), the corresponding proton acceptor, constitute a conjugate acid-base pair, related by the reversible reaction.

- Each acid has a characteristic tendency to lose its proton in an aqueous solution.
- The stronger the acid, the greater its tendency to lose the proton.
- The tendency of any acid (HA) to lose a proton and forming the conjugate base (A^-) is defined by equilibrium constant (K) for the reversible reaction.
- Equilibrium constant for an ionization reaction is known as ionization or dissociation constant, often designated as Ka (dissociation constant of acid).

$$K_{eq} = \frac{\left[H^+\right]\left[A^-\right]}{\left[HA\right]} = Ka$$

- Analogous to pH, pKa is defined by the equation

$$pK_a = \log\frac{1}{K_a} = -\log K_a$$

- The stronger the tendency to dissociate/donate a proton, the stronger is the acid and lower is its pKa.
- The pKa of a weak acid can be determined quite easily by titration reaction.

Titration

Also known as titrimetry, it is a chemical (qualitative and volumetric) analysis used to calculate the concentration of a given analyte (e.g., acid or base) in a mixture. It is an important volumetric analysis in the field of analytical chemistry.

Titration Curve

Titration curve reveals the pKa of weak acid. Titration is performed to determine the amount of base (of known strength) is required to neutralize a given (acidic) solution, so as to calculate the strength of the solution. A measured volume of the acid is titrated with a strong base (of known strength) like NaOH (or vice versa). The base is added in small increments until the acid in the solution is neutralized, as determined with an indicator dye or using a pH meter. The concentration of the acid in the original solution can be calculated from the volume and concentration of the base added. A plot of the pH values against the amount of acid added, represents the titration curve which reveals the pKa of the weak acid (Figure 10.4).

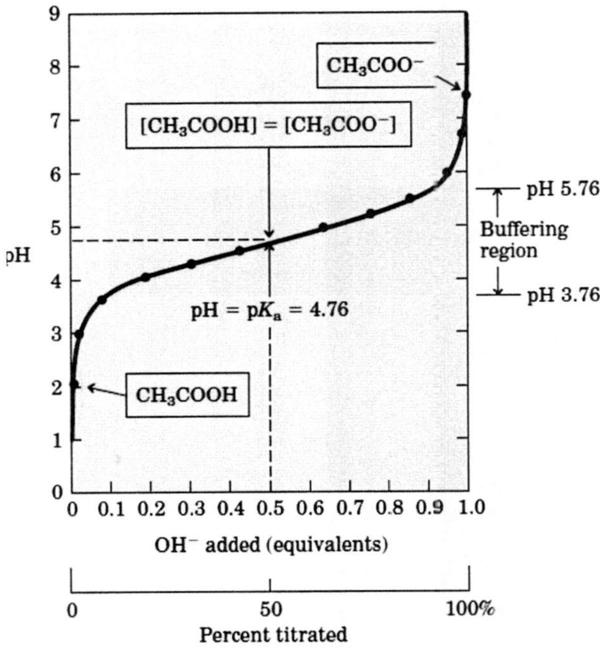

Fig. 10.4: Titration curve

For example: In a titration of 0.1 M solution of acetic acid with 0.1 M NaOH at 25 °C, NaOH is added gradually, the OH^- combines with the free H^+ in the solution to form H_2O to an extent that satisfies the equilibrium relationship. As the free H^+ is removed, acetic acid dissociates to satisfy the own equilibrium constant. As the titration proceeds, more and more acetic acid ionizes and forms acetate ion. At a point of titration when exactly 0.5 equivalent of NaOH has been added into the acetic acid, one-half of the original acetic acid gets dissociated. At this midpoint, the concentration of proton donor (CH3COOH) is equal to that of the proton acceptor ($CH3COO^-$). At this point, a very important relationship holds wherein the pH of the equimolar solution of acetic acid and acetate is equal to the pKa (4.76) of acetic acid. One of the most important points about the equimolar mixture of weak acid and its conjugate base is that it acts as a buffer within the range ±1of its pKa value (shaded zone in Figure 10.4).

Type of Buffer

There are two types of buffers, i.e., acid buffer and base buffer.

Acid Buffer: A buffer that contains large quantities of a weak acid, and its salt with a strong base, is called an acid buffer. Such buffers have the pH below 7. One of the examples is the mixture of CH_3COOH and CH_3COONa.

Basic Buffer: A buffer solution which contains relatively large quantities of a weak base and its salt with a strong acid is called a basic/simple buffer. Such buffers have the pH more than 7 at 298 K. One of the examples is the mixture of NH_4OH and NH_4Cl.

Mechanism of Action of Acidic Buffers

In a buffer of CH_3COOH (weak electrolyte) and CH_3COONa (strong electrolyte), there will be a large concentration of Na^+ ions, CH_3COONa^- ions, and un- dissociated CH_3COOH molecules. ***When a strong acid like (HCl) is added*** in $CH_3COOH–CH_3COONa$ buffer, the changes that will occur may be represented as:

$$CH_3COONa \rightarrow Na^+ + CH_3COO^- \qquad H^+ + Cl^- \leftarrow HCl$$

$$CH_3COOH$$

The hydrogen ions (H^+) yielded by the HCl will be quickly removed as unionized acetic acid, and the hydrogen ion concentration will, therefore, be only slightly affected, because the produced acetic acid is weaker compared to the added HCl.

When a strong base like (NaOH) is added in $CH_3COOH–CH_3COONa$ buffer, the changes that will occur may be represented as:

$$CH_3COOH \rightarrow CH_3COO^- + H^+ \qquad OH^- + Na^+ \leftarrow NaOH$$

$$H_2O$$

The hydroxyl ions (OH^-) yielded by the NaOH will be removed in the form of water. The supply of hydrogen ions (H^+) needed for this purpose will be provided by the dissociation of acetic acid.

Mechanism of Action of Basic Buffers

In a buffering system of NH_4OH (weak electrolyte) and NH_4Cl (strong electrolyte), there will be a large concentration of NH4$^+$ ions, Cl^- ions, and un- dissociated NH_4OH molecules.

When a strong acid (HCl) is added in NH_4OH–NH_4Cl buffer, the changes that will occur can be represented as:

$$NH_4OH \rightarrow NH_4^+ + OH^- \qquad H^+ + Cl^- \leftarrow HCl$$

$$H_2O$$

The hydrogen ions (H^+) yielded by the HCl will be removed as water. The supply of OH^- needed for this purpose will be provided by the ammonium hydroxide.

When a strong base (NaOH) is added in NH_4OH–NH_4Cl buffer, the changes that will occur can be represented as:

$$NH_4Cl \rightarrow Cl^- + NH_4^+ \qquad OH^- + Na^+ \rightarrow NaOH$$

$$NH_4OH$$

The hydroxyl ions (OH^-) yielded by the NaOH will be removed as unionized ammonium hydroxide, and the pH of solution will be slightly affected.

Preparation of Buffers

A buffer solution is a mixture of weak acid and its corresponding conjugate bases or a weak base with its corresponding conjugate acids. If the dissociation constant of the acid (pKa) and of the base (pKb) is known, a buffer can be prepared by calculating the salt-acid or the salt-base ratio. There can be three different methods to prepare a buffer.

First method: *By titration* of one of the two components of the buffer with strong base/acid. Prepare a buffer using a weak acid (e.g., acetic acid) and its salt (CH_3COONa) by adding a strong base (e.g., NaOH) until the desired pH is obtained.

If another component of the buffer is available (e.g., sodium acetate), a strong acid (e.g., HCl) can be added until the desired pH is obtained.

Second method: *Using pKa value of the buffer*, calculate the amounts (in moles) of acid/salt or base/salt need to be present in the buffer at the desired pH. If both the forms (i.e., the acid and the salt) are available, convert the amount from moles to grams (using the molecular weight of the component) and then weigh the required amounts. Convert moles to volume, if the stock is available in the liquid form.

Third method: Using buffer table find out the amount of acid/salt or base/ salt required to prepare the buffer of desired pH. Dissolve the components in slightly lesser volume (less than the required final volume of buffer solution) of water. Check the pH and make it up if necessary. Make up the final volume by adding distilled water.

Let's Check Your Understanding

1. Which of the following will not function as a buffer?

   a. NaOH + HCOONa   b. Borax + Boric acid

   c. $NaH_2PO_4 + Na_2HPO_4$ d. $NH_4Cl + NH_4OH$

2. The two important biological buffers are ________ and __________.

3. Which one of the following is not an acidic buffer?

   a. H_2CO_3 and Na_2CO_3 b. CH_3COOH and CH_3COONa

   c. $HClO_4$ and $NaClO_4$ d. H_3PO_4 and Na_3PO_4

4. A solution with pH= 0 is

   a. acidic   b. basic

   c. amphoteric   d. neutral

5. Which of the following is the strongest conjugate base?

   a. Cl^- b. SO_4^{2-}

   c. CH3COO$^-$ d. NO3$^-$

6. What will be the pH of 0.1N NaOH solution.

   a. 13   b. 10

   c. 12   d. 11

7. Buffer solution resists any drastic change in pH because ______

   a. acid and alkali in the buffer solution is protected from attack by other ions.

   b. it gives unionised acid or base on reaction with added acid or alkali.

   c. fixes the pH value.

   d. it contains large excess of H^+ or OH^- ions.

8. Considering a solution which contains 0.1 M CH_3COOH and 0.2 M CH_3COONa. Which of the following statements is true?

   a. If a small amount of NaOH is added, the pH decreases very slightly.

   b. If NaOH is added, the OH^- ions react with the CH3COO$^-$ ions.

   c. If a small amount of HCl is added, the pH decreases very slightly.

   d. If HCl is added, the H^+ ions react with CH_3COO^- ion.

9. A solution with a pH greater than 7 is considered acidic.

True / False

10. The pH of a solution can be accurately measured using litmus paper.

True / False

Answers

1. a
2. Phosphate buffer and Bicarbonate buffer
3. c
4. a
5. c
6. a
7. b
8. c
9. False
10. False

11

Handling of Acid/Base and Titration in Laboratory

Introduction

Use of acid/base is common in laboratories for experiments like titration and proper safety in their handling laboratory is necessary. In this chapter we will briefly discuss about the safety measures necessary to be adopted while using acids/bases in laboratory for any experiment.

Handling of Strong Acid/Base

There are four major points that need to be taken care of while working with strong acids or bases that might be quite harmful if any accident happens.

1. Protect Yourself

Wear a lab-coat and acid-resistant gloves: Always use lab-coat/apron and make it sure that the sleeves properly cover your arms. More importantly, use gloves made of acid-resistant materials and also wear shoes to protect your feet while wearing long-pants/jeans. Female with long hair, must tie her hair properly (with rubber band) to avoid any disturbance caused by appearance in between.

Wear safety goggles: It is important to wear safety goggles to protect your eyes using larger and fitting goggles that cover front as well as sides of your eyes. Always use properly fitting and adjustable goggles because smaller/ loose goggles may not protect the areas around your eyes against accidental acid spills and it may fall down from your face, if it is too loose.

Locate emergency wash area: There must be an emergency washing area nearby the work place or laboratory containing lab spill management kits. All the workers in the laboratory must be aware of using the kits and emergency plans.

Neutralize acid: Don't panic even if an acid or chemical spill happens. Stay calm and act as quickly as possible. In case an acid spills, neutralize it by using sodium bicarbonate so that they can be cleaned and disposed safely. If acid comes in contact of skin, wash it off under running tap water but do not use/apply sodium bicarbonate on skin.

2. Use Appropriate Lab Equipment/Practice

Use acid-compatible containers: Always use acid-resistant plastic wares (for example, those made of LPDE and PVC). However, different acids/chemicals have different impacts on these materials; therefore, it is important to choose suitable material for storage and handling of acids as well as other chemicals. For example, chromic acid (10%–50%) can be safely stored in low density polyethylene (LDPE) container while acetic acid (50%) should be stored in a container made of polypropylene (PP). A chemical compatibility chart can be made available in a laboratory to avoid any misuse/accident.

Do not use leaking or damaged containers: If you notice any container to be damaged or leaking, remove/replace it.

Use a fume hood: For safety purpose, it is important to use fume hood for handling of volatile/hazardous chemicals.

Store acids separately from other chemicals: **Always store acids separately from other chemicals**, particularly from volatile organic chemicals and label them properly for convenience in searching/identification of the container. Separate cabinet for corrosive chemicals/acid can be made keep them safely.

3. Use Proper Method for Preparing Solution

Follow the MSDS: Before using any acid/hazardous material in a laboratory, make yourself aware about the chemical using/reading "Material Safety Data Sheet" (MSDS) also known as product safety data sheet (PSDS) and now as safety data sheet (SDS), which provides necessary precautions, safety measures, appropriate protocols and other relevant information about the chemical.

Use appropriate equipment while handling acids: Always use acid-safe containers (made of borosilicate or pyrex) while using/measuring concentrated or distilled acid. Use appropriate device to handle/transfer acids, as mouth pipetting needs to be avoided in a laboratory.

Always add acid to water: Never add water to acid as it can produce acidic steam and may cause spills and injuries. While making dilutions, always add acid to water because spilling of water will not be a problematic. Once the dilution/ solution cools down to room temperature, add more water to make up the final volume.

4. Be Aware of Emergency Handling Procedures

Utilize emergency showers or eyewash stations: If you get exposed to any acidic material, then you must immediately proceed to emergency washing area or showers (in case the whole body gets exposed). Continuously flush the contacted area under running water for about 15 minutes. Remove all the contaminated clothing.

Cleaning the spill: If the spill can be handled by the staff safely, then it should be cleaned by the laboratory staff as per the laboratory guidelines. However, if spill is larger, appropriate arrangement should be made to clean it taking help of the experts.

Evacuate the area and rush to medical help, if needed: In case of severe spills, all members/staff of the laboratory will need to evacuate the room. Inform about the accident to the biosafety committee members of the institution. Depending on the severity of accidents (any injury to eye, inhalation of acid fumes, etc.), seek medical consultation by visiting nearest doctor/hospital.

Titration

Titration is a procedure wherein a solution of known concentration (titrant) is added to a known volume of another solution (titrant or analyte) in order to determine its concentration. Many of the titrations are based on acid-base neutralization reaction. To perform an acid-base titration, we need to have a way to visually detect the neutralization point.

The titrant is added slowly using a burette to a known the volume until the reaction (neutralization) is complete. Knowing the volume of titrant added allows to determine the concentration of the unknown analyte. Acid-base titrations are monitored by the change of pH as titration progresses. An indicator can also be used which indicates neutralization point by colour change. A commonly used indicator in strong acid – strong base titrations is phenolphthalein. When a few drops of phenolphthalein dye, a man-made visual indicator, is added to an acidic solution it turns colourless, but the solution becomes brilliant pink as the solution turns to basic.

Steps in a Titration

1. A measured volume of an acid of unknown strength is taken in an Erlenmeyer flask.
2. A few drops of an indicator can be added to the analyte (acid) and mixed by swirling.
3. A burette is filled with titrant solution (base) of known strength.
4. Stopcock of the burette is opened and the titrant is dropped slowly into the analyte while continuously swirling the flask.
5. The pH of analyte solution is recorded using a pH meter with increasing volume of titrant added.
6. For determining the strength of analyte, stopcock of burette is closed at the moment the pH of analyte solution becomes 7 (neutralized).

Preparation of Titration Curve

Titration curve is prepared by plotting the pH value of the test (analyte) solution on Y-axis recorded with the addition of increasing volume of titrant (standard solution) on X-axis.

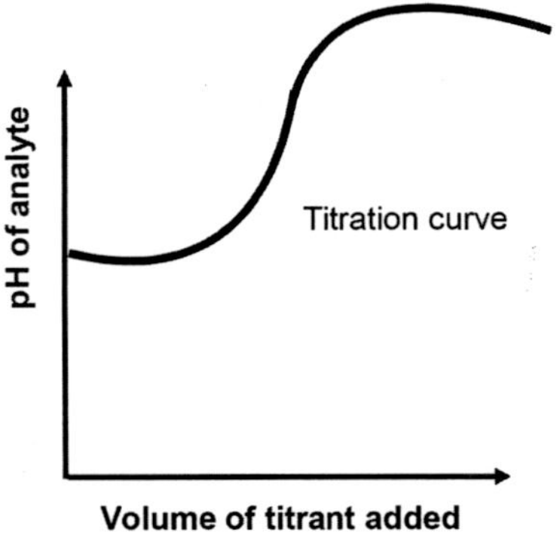

Fig. 11.1: Titration curve

Let's Check your Understanding

1. Which of the following is used as an indicator in the titration of a strong acid and a weak base?

   a. Methyl orange   b. Phenolphthalein
   c. Thymol blue   d. Fluorescein

2. Find the concentration of HCl, if 10 ml of 0.5 M $Ca(OH)_2$ is required to titrate 50 ml of HCl.

   a. 0.5 M   b. 0.1 M
   c. 5.0 M   d. 1.0 M

3. Which of the following is used as an indicator in the titration of a weak acid and a strong base?

   a. Methyl orange   b. Bromothymol blue
   c. Phenolphthalein   d. Methyl red

4. A difference between strong and weak acid is ---------.

   a. proton donation and electron acceptance.
   b. complete and partial ionization.
   c. negative and positive Ph.
   d. presence and absence of halogen ions.

5. A separate cabinet must be used for storage of acids in a laboratory.

   True / False

6. Always put water into acid while making solution/dilution.

   True / False

7. Acid-base titrations are monitored by the change in pH as titration progresses.

   True / False

8. If your skin gets touched with acid, wash it with running water and put sodium bicarbonate on your skin.

   True / False

9. In a titration, the endpoint is the exact point at which the chemical reaction between the titrant and the analyte is complete.

   True / False

10. Full form of MSDS

Answers

1. a
2. b
3. c
4. b
5. True
6. False
7. True
8. False
9. False
10. Material Safety Data Sheet.

12

Preparation of Medium and Sterilization

Introduction

Different types of culture media are used for culturing microbes and plant tissues. Preparing as well as sterilizing them properly following appropriate method with safety for the desired purpose are important parts of research work. In this chapter, we will discuss briefly about different types of media, their composition, preparation and sterilization in laboratory for microbial or plant tissue culture.

Culture Medium

Ingredients mixed to prepare solid, liquid or semi-solid substance to support the growth of microorganism or plant tissue *in vitro* (outside the body) is called culture medium. Frequently, it becomes important to grow microorganism or plant/animal tissue *in vitro* for any or all of the following purposes:

- To study characteristic features or properties of microorganisms.
- To identify the causal organism(s) of infection in a clinical sample, so that appropriate remedy can be determined.
- To culture/multiply microorganism or plant/animal tissues *in vitro* for research or commercial purposes.
- To prepare/produce biological molecules, vaccines, antigens, proteins, metabolites, etc. *in vitro* for research or industrial purposes.

Important Ingredients/Factors of Culture Medium

- Water,
- Source of energy,
- Source of carbon,
- Source of nitrogen,

- Minerals,
- Special growth factor.

Types of Culture Media

Classification Based on Physical State of the Medium

1. *Solid medium*: The medium prepared using solidifying agent like Agar (a most commonly used solidifying agent), Phytagel, etc. and appears in solid form on cooling or when used. An example of solid medium is Luria agar. One of the gelling agents commonly used for the purpose is Agar agar. It is a golden-yellow granular powder, prepared from seaweeds, which melts at 98 °C and sets/solidifies at 42 °C.
2. *Semi-solid medium*: Such medium is soft (semi-solid) and prepared by using lesser amount of solidifying agent. These media are useful in demonstrating bacterial motility and separating motile from nonmotile strains.
3. *Liquid medium*: Such a medium is also referred as broth and used for bacterial growth or cell culture. An example of liquid medium is Nutrient broth or Luria broth.

Classification Based on Ingredients of the Medium

1. *Simple medium*: A medium whose exact composition is known/well-defined is known as simple medium. One of the examples of such medium is Murashige and Skoog (MS) medium.
2. *Complex medium*: A medium whose exact composition is not known/ defined is known as complex medium. One of the examples of such medium is Blood agar medium, as its composition/ exact composition is not known.
3. *Synthetic or defined medium*: Such a medium is prepared for research purpose and components of the medium are known. One of the examples of such medium is Peptone yeast extract medium.
4. *Special medium*: Such medium is prepared for specific usage in research work. These media include Enriched medium, Selective medium, Differential medium, Transport mediu,, Anaerobic medium, etc.

Points to be Considered While Preparing a Medium

A medium can be distributed in culture vessels before autoclaving or the medium and vessels can be autoclaved before they are autoclaved. For preparing solid medium, gelling/solidifying agent is added to the medium in appropriate concentration before autoclaving. Agar medium is often autoclaved together in larger (500 ml) volume and subsequently poured into sterilized containers under a laminar flow hood. Heat-labile components (such as L-glutamine and gibberellic acid, antibiotics, etc.) should be filter sterilized and added to the autoclaved medium after cooling down but before solidification. For filter sterilization, the chemical is first dissolved in a lesser volume of the solvent (generally water) and then final/desired volume is made up with the solvent. Filter sterilization is used for physical removal of microorganisms from a thermally sensitive solution. Filtration is carried out using a sterile filter of appropriate size (0.2 μm) membrane and stored in a sterile container while carrying out the filtration process under sterile conditions (laminar air flow hood).

Precautionary Steps in Preparing a Medium

1. Follow step-by-step procedure to prepare a medium, which ensure identical composition of the medium every time it is prepared and it avoid any mistake.

2. Add each component according to a list of ingredients and mark it on the list once it is added. Stock solutions of ingredients/constituents of the medium must be prepared by completely dissolving the ingredients. Before using stock solution to prepare the medium, ensure that there is no precipitation in the stock solution.

3. Measure/weigh each constituent chemicals carefully/precisely using pipette/graduated cylinder or appropriate weighing balance. To prepare am aqueous stock solution, add the constituent chemicals in a smaller volume of water one-by-one agter completely dissolving it.

4. While preparing a medium, add the constituent stock solutions in a small volume of water taken in a beaker. Adjust the pH of medium to the required value using 1.0 M HCl or NaOH by adding it drop-wise and measuring it with the help of pH meter while stirring (using a magnetic stirrer). Make up the final volume of the medium using a graduated measuring cylinder or graduated beaker and then add the gelling agent.

5. Autoclave the medium at 121 °C at 105 kPa for 18 minutes. However, ensure that the container filled with medium for autoclaving must have sufficient air space (not filled more than 70%) to prevent any boiling

off the medium from container. In case more than one medium are autoclaved in different container, label the containers with the name of medium. Also, use autoclave indicator tape to be informed about proper autoclaving of the medium.

6. Be careful while taking out the sterilized medium from autoclave, particularly when opening the autoclave (to save your face from the steam suddenly coming out of the autoclave when lead is opened). Use thermal cloves to take out the container from the autoclave, particularly when you are in a hurry (do not have sufficient time to let the medium cool down to bearable warmth).
7. Allow the medium to cool down to bearable warmth, for which the container should be swirled gently (without opening the container) to get it cooled.
8. When the medium is sufficiently cooled, add the thermo-labile component (filet sterilized), if any. Dispense the medium into sterile containers/Petri plates under aseptic (laminar air flow hood) conditions.
9. Keep the Petri plates half-open after pouring the medium until it solidifies (inside the running laminar flow hood) to avoid condensation of water droplets on the lead. Label the containers with the name of the medium as well as the date of preparation before storing the containers for later use.

Composition of Some of the Medium (Per liter)

1. Lactose broth (Seeley and Vademark, 1970)

Yeast extract	: 03.00 g
Peptone	: 05.00 g
Lactose	: 05.00 g
Distilled water	: to 1000 mL
pH	: 6.5

2. MRS (de Man, Rogosa and Sharpe) agar medium (De man et al., 1960)

Peptone	: 10.00 g
Beef extract	: 10.00 g
Yeast extract	: 05.00 g
K_2HPO_4	: 02.00 g

Triammonium citrate	:	02.00 g
Glucose	:	20.00 g
$MgSO_4$	:	00.58 g
$MnSO_4$	:	00.25 g
Sodium acetate	:	05.00 g
Agar	:	18.00 g
Distilled water	:	to 1000 mL
pH	:	6.2

3. Mannitol salt agar medium (Chapman, 1945)

NaCl	:	75.00 g
Proteose peptone	:	10.00 g
Mannitol	:	10.00 g
Beef extract	:	1.00 g
Phenol red	:	0.025 g
Agar	:	20.00 g
Distilled water	:	to 1000 mL
pH	:	7.4

4. Malt extract agar medium (Thom and Church, 1926)

Malt extract	:	20.0 g
Dextrose	:	20.0 g
Peptone	:	6.0 g
Agar	:	15.0 g
Distilled water	:	to 1000 mL
pH	:	5.5

5. Martin's rose Bengal agar (Martin, 1950)

Glucose	:	10.00 g
Peptone	:	5.00 g
Yeast extract	:	0.50 g

K_2HPO_4	:	0.50 g
KH_2PO_4	:	0.50 g
$MgSO_4 .7H_2O$	:	0.50 g
Rose Bengal	:	30.00 mg
Agar	:	15.00 g
Distilled water	:	to 1000 mL
Streptomycin sulphate	:	30.00 mg

6. Methyl Red and Voges-Proskauer medium (Clarke and Kirner, 1941)

Peptone	:	7.00 g
KH_2PO_4	:	5.00 g
Dextrose	:	5.00 g
Distilled water	:	to 1000 mL
pH	:	7.0

7. Nutrient agar medium (Anon, 1957)

Beef extract	:	3.00 g
Peptone	:	5.00 g
Agar	:	17.00 g
Distilled water	:	to 1000 mL
pH	:	7.0

8. Nutrient gelatin medium (Ewing, 1962)

Beef extract	:	3.00 g
Peptone	:	5.00 g
Gelatin	:	120.00 g
Distilled water	:	to 1000 mL
pH	:	7.0

9. Skim milk agar medium (Aneja, 2012)

Pancreatic casein digest : 05.00 g

Yeast extract : 02.50 g

Powdered non-fat milk : 100.00 g

Glucose : 01.00 g

Agar : 15.00 g

Distilled water : to 1000 mL

pH : 7.0

10. Starch agar medium (Aneja, 2012)

Beef extract : 3.00 g

Soluble starch : 10.00g

Agar : 12.00g

Distilled water : to 1000 mL

pH : 7.5

11. Sucrose agar medium (Dolan and Morgan, 1978)

Sucrose : 150.00 g

Bactro-peptone : 5.00 g

Yeast extract : 5.00 g

K_2HPO_4 : 15.00 g

$MnCl_4\ H_2O$: 0.01g

NaCl : 0.01g

$CaCl_2$: 0.05g

Agar : 15.00g

Distilled water : to 1000 mL

pH : 7

12. Tributyrin agar medium (for lipid hydrolysis test) (Aneja, 2012)

Beef extract : 3.00 g

Peptone : 5.00 g

Agar : 15.00 g

Tributyrin oil	:	10.00 mL
Distilled water	:	to 1000 mL
pH	:	6.0

Sterilization of Medium

Microorganisms are omnipresent and many of them cause contamination, infection, and decay. Hence,, it becomes necessary to remove/destroy them. Sterilization process should be simple, effective, and relatively easier. Sterilization of an article, surface, or medium is to make it free from microorganisms present either in the vegetative or spore form. Moreover, in case of thermo-labile materials, they are disinfected by removal of microorganisms (e.g., by filter sterilization, instead of heat sterilization).

Different Methods of Sterilization

1. ***Dry heat sterilization***: Empty glass ware (culture vessels, petri plates etc.) certain plasticware (Teflon), Metallic instruments (scalpels, forceps, needles, etc.) Aluminium foils can be sterilized by exposure to hot dry air at 160–180 °C for 2–4 hour in a hot air oven. However, the items should be properly sealed before sterilization.

2. ***Wet heat/vapour sterilization***: In this method of sterilization, water vapour under high pressure is utilized to kill germs/microbes by exposing them to super-heated steam in an autoclave. Normally the tissue culture media (in glass container covered with cotton plugs, Aluminium, foil, plastic caps) are autoclaved at a pressure of 15–18 psi at 121 °C for 15–-20 minutes. The advantages of using wet heat/autoclave for sterilization are the speed, efficiency, simplicity and destruction of microbes, while disadvantages are change in pH by 0.3 – 0.5 units.

3. ***Filter sterilization***: Vitamins, amino acids, plant extracts, hormones, growth regulators are thermo-labile compounds may get destroyed during autoclaving. Hence, such chemicals are filter sterilized by passing through a bacterial proof membrane under positive pressure. A Millipore or Seitz filter with a pore size ≤0.2 μm is generally used for filter sterilization. Filter sterilization must be carried out under aseptic conditions/laminar air flow cabinet.

4. ***Ultra violet sterilization***: UV light is used to sterilize interior portion of inoculation chamber/laminar air flow cabinet and atmospheric contamination. Materials like disposable plasticware, surface area,

etc. used for tissue culture are sterilized using UV rays to minimize contaminates.

5. ***Flame sterilization***: Metallic instruments like forceps, scalpels, needle, and spatula are sterilized by dipping in 95% ethanol followed by flaming sterilization and cooling. Autoclaving of metallic instrument is generally avoided as they get rusted and become blunt. Generally mouth of culture vessels need to be sterilized by exposing to flame prior to inoculation.

6. ***Chemical / surface sterilization***: Most of the living/ explant materials used for tissue culture are surface sterilized using an appropriate chemical (e.g., $HgCl_2$, Mercuric chloride, etc.) before transferring to the nutrient medium for culture. Chemical surface sterilization is generally performed using Mercuric chloride (0.1% for 3–-10 minutes, Calcium hypochlorite 5% for 20 minutes, Sodium hypochlorite 0.5–1.0% for 15 minutes, Bromine water 1.0% for 2–10 minutes.

The surfaces that cannot be sterilized by any other method (for example parts of laminar air flow cabinet, palm of the researcher, etc.), are sterilized by wiping them thoroughly with 70% ethanol and allowed to dry.

Safety Measures While Working with Autoclave

1. Always wear lab coat with full arms, PPE or heat-tolerant gloves, as per the requirements to protect yourself.

2. Prepare the autoclave by checking the gasket on lead/door against any cracks (it should be smooth without any break).

3. Do not autoclave flammable, combustible, reactive, corrosive, toxic, or radioactive materials. Do not autoclave cracked/compromised glassware.

4. For autoclaving liquids, leave cap of the container loose or cover with aluminium foil to allow passage of steam/gas to avoid explosion. For bagged items, loosely tie the bag leaving opening for steam to go inside the bag.

5. Never place autoclavable items directly on the autoclave bottom or floor. Don't overload/pack an autoclave and keep sufficient space for steam formation.

6. Follow the instructions mentioned in the user manual provided by the manufacturer and abide by the laboratory standard operating procedure (SOP) for using the autoclave. Do not open the lead/door of autoclave in

a hurry, immediately after sterilization time is over, releasing pressure or during the run. If necessary, abort the autoclaving and wait until the pressure inside the autoclave chamber is completely released.

7. When pressure inside the autoclave is released, make sure that chamber temperature has dropped down to just above room temperature before opening the lead. Keep your face away from the steam coming out of the lead/door. Keep the lead/door of autoclave open for 10 minutes and then carefully remove items one by one. Do not agitate/swirl autoclaved/hot containers, as boiling or superheated liquids can explode if agitated rapidly.

Let's Check Your Understanding

1. Liquid medium is also called ....
2. Autoclaving may changes the pH by 0.3 – 0.5 units.

   True / False
3. You can open the autoclave on the run.

   True / False
4. Name the most commonly used solidifying agent.
5. Name some of the chemicals used for surface sterilization of explant.
6. What is the main purpose of adding a gelling agent like agar to a culture medium?
   a. To increase the nutritional content of the medium
   b. To solidify the medium
   c. To sterilize the medium
   d. To provide a source of carbon
7. Which of the following is an example of a simple medium whose exact composition is known?
   a. Blood agar medium
   b. Nutrient broth
   c. Murashige and Skoog (MS) medium
   d. Peptone yeast extract medium
8. What precaution should be taken when autoclaving liquids to avoid explosion?
   a. Tighten the cap of the container securely
   b. Leave the cap of the container loose or cover with aluminum foil
   c. Place the container directly on the autoclave floor
   d. Fill the container completely to minimize air space
9. Which sterilization method is suitable for heat-sensitive materials like vitamins and hormones?
   a. Dry heat sterilization
   b. Wet heat/vapour sterilization
   c. Filter sterilization
   d. Flame sterilization

10. What is the recommended maximum percentage of amaranth flour to be added to wheat bread to maintain product quality and nutritional benefits?

    a. 10%
    b. 15%
    c. 20%
    d. 25%

Answers

1 Broth
2. True
3. False
4. Agar-agar
5. Mercuric chloride, Calcium hypochlorite, Sodium hypochlorite, Bromine water, etc.
6. b
7. c
8. b
9. c
10. c

13

Plant Tissue Culture

Introduction

Tissue culture refers to the technique of growing cells, tissues, or organs under controlled, artificial conditions. This allows propagation of tissues from a small sample of cells/tissues under sterile conditions. Plant tissue culture is widely used for various purposes, including the production of disease-free plants, rapid multiplication of plants with desirable traits, genetic modification, conservation of rare or endangered species, etc. Tissue culture is a valuable technique because it enables consistent and large-scale production of plants, ensures genetic uniformity, and allows manipulation of plant biology at cellular level. In this chapter, we will discuss about the major types of tissue culture techniques and their advantages, disadvantages, as well as applications in brief.

Different Types of Tissue Culture

Callus culture involves induction and growing of unorganized, undifferentiated cell masses from an explant on a nutrient/culture medium. The part of plant used to induce callus is called explant which can be tissue, organ or part of a plant.

Organ culture is used to maintain or grow an entire organ or a part of organ *in vitro*. It preserves the structure and function of the organ for studies or transplantation.

Embryo culture involves isolating and growing embryos from seeds or ovules in a nutrient medium. It is used to bypass seed dormancy or rescue embryos from hybrid crosses that may not develop naturally.

Cell suspension culture involves growing dispersed cells or small cell clusters in a liquid medium under continuous shaking. This is used for large-scale production of cells, secondary metabolites, and studying cell biology.

Protoplast culture uses cell without cell wall (known as protoplast) in a culture medium to regenerate the cell and then the whole plant. This technique is used in genetic engineering and somatic hybridization experiments.

Anther/Pollen culture is used to produce haploid plant, which can be doubled to generate homozygous line. This is a very useful technique for genetic studies and plant breeding.

1. Callus Culture

Callus culture involves growth and multiplication of undifferentiated cells (callus) induced from an explant on a nutrient medium under aseptic conditions. This technique is useful for research, breeding, and genetic modification experiments in plant. The steps involved in callus culture are as follows.

Selection of explant: A suitable young and actively growing plant tissue (e.g., leaf, stem, root) from a healthy plant is preferred as explant.

Sterilization of explant: The explant is surface sterilized to eliminate contaminants commonly using agents like 70% ethanol and sodium hypochlorite (1–2% solution). The explant is thoroughly rinsed with sterile distilled water to remove any traces/ of sterilizing agents.

Preparation of culture medium: Appropriate nutrient medium such as Murashige and Skoog (MS) medium, supplemented with the necessary plant growth regulators (e.g., auxins and cytokinins) is prepared for culturing the explant *in vitro*. For callus induction, in general, medium containing a combination of 2,4-D (2,4-Dichlorophenoxyacetic acid, an auxin) and kinetin (a cytokinin) is used.

Inoculation of medium: The sterilized explant is cut into small pieces (about 1 cm^2) and placed on the medium in sterile container (Petri dish or culture bottle/jar).

Incubation

The culture is incubated in a controlled condition with appropriate temperature (usually 25 °C) and light (in dark to induce callus).

Callus induction/formation: After a few days/weeks of incubation of explant on callus induction medium (Figure 13.1), initiation of callus formation can be seen. The explant cells will proliferate and form a mass of undifferentiated cells.

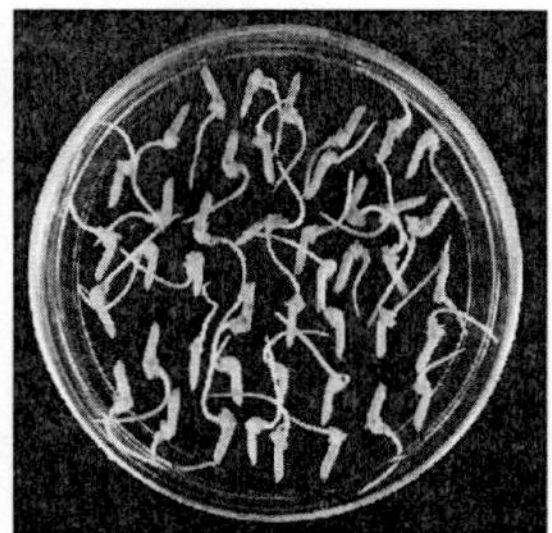

Fig. 13.1: Inoculation of explant (mature seeds) for callus

Subculturing: Once a sufficient amount of callus is formed on the explant, calli are transfer (subcultured) to a fresh medium for continued growth (proliferation or multiplication) of calli to be utilized in the subsequent steps.

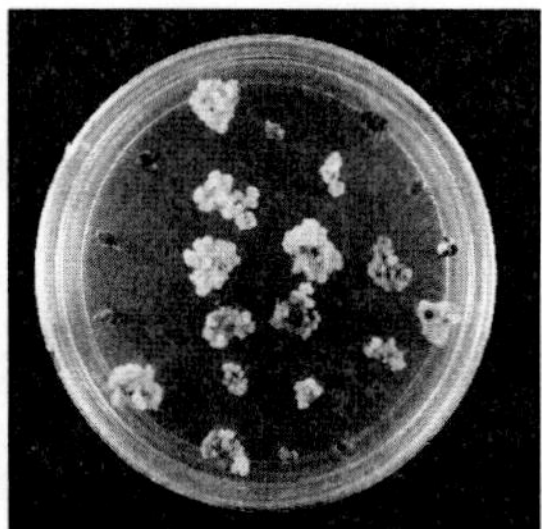

Fig. 13.2: Callus proliferation on subculturing.

Regeneration: To regenerate plants, calli are transferred to regeneration medium containing different plant growth regulators (phytohormones) that promote organogenesis (shoot and root or both).

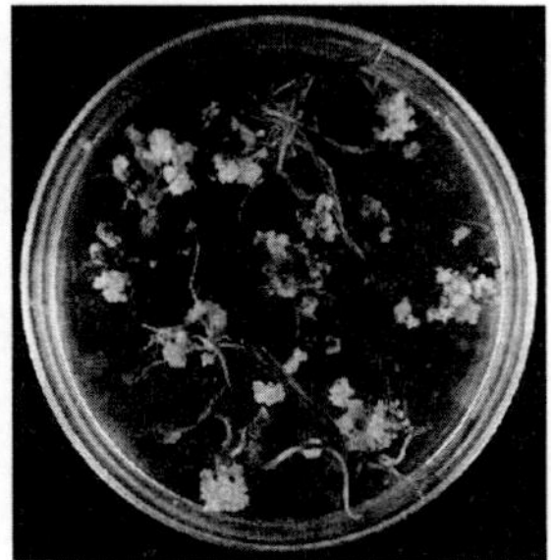

Fig. 13.3: Regeneration of plants from callus.

Advantages of Callus Culture

Genetic manipulation: Provides an easier platform for genetic transformation/ gene editing for basic/applied studies on plants.

Propagation: Enables rapid multiplication of plants with desirable traits; however with a risk of somaclonal variations.

Secondary metabolite production: Can be used for the production of valuable secondary metabolites being formed in the undifferentiated cells in large quantity.

Studies on plant development: Offers insights into plant cell differentiation and developmental processes.

Disadvantages of Callus Culture

Genetic variability: Callus culture may exhibit somaclonal variations (induced during multiplication of calli) leading to genetic instability and unwanted variation.

Labour-intensive: Requires careful maintenance through frequent subculturing of calli for faster multiplication, which makes it a labour-intensive and time-consuming process.

Costly maintenance: Maintenance of callus involves careful handling callus under sterile conditions to avoid contamination of culture. This leads to higher cist of maintaining callu culture for longer period.

Risk of contamination: The culture is highly susceptible to microbial contamination, which requires skilled personal to handle the culture.

Regeneration challenges: Not all the plant species or tissues are equally amenable to regeneration of plant from callus, which poses a challenge in tissue culture of the plant species.

2. Organ Culture

Organ culture involves the growth and maintenance of an organ or a part of it in controlled, artificial conditions. This preserves the architecture and function of the organ, allowing for detailed studies on organ development, function, and response to various treatments. The steps involved in organ culture process are as mentioned below.

Selection of explant: A healthy part of plant/organ (e.g. shoot tip) is selected for organ culture.

Sterilization of explant: The explant is surface sterilized to eliminate contaminants with imposing much damage to the explant. The commonly used sterilization process a treatment with 70% ethanol for a short (90–120 seconds) duration followed by treatment with diluted solution of sodium hypochlorite

(0.1% for 10 minutes). The explant is rinsed thoroughly with sterile distilled water to remove traces of the sterilizing agents.

Preparation of culture medium: An appropriate nutrient medium (such as Murashige and Skoog) is prepared and sterilized. It is supplemented with necessary plant growth regulators. Typically, a combination of 6-benzylaminopurine (BAP) and indole-3-acetic acid (IAA) is used for shoot tip culture.

Inoculation: The sterilized explants are placed on the culture medium in Petri plates or jars under sterile conditions.

Incubation: The culture is incubated under controlled conditions, typically at 25 °C with a 16 hour photoperiod (light and dark cycle).

Growth and maintenance: Shoot tips are maintained for growth and development of shoots and roots depending on the composition of phytohormones used in the culture medium. The growing culture is subcultured to fresh medium periodically to maintain nutrient availability to the growing plants for continuous growth until optimum growth is attained.

Advantages of Organ Culture

Structural/functional preservation: It maintains the architecture and function of the organ, allowing rapid studies needed on intact organ.

Developmental studies: Enables detailed observation of organ development, differentiation, and morphogenesis.

Testing and screening: Facilitate the testing of various compounds (e.g., hormones, drugs) for their effects on the organ, useful in both plant and animal research.

Pathogenic study: Allows studies on organ-specific pathogen interactions under controlled environment.

Genetic stability: This typically exhibits lesser genetic variability compared to that on callus culture, as it maintains the genetic integrity.

Disadvantages of Organ Culture

Technical challenges: It requires precise handling and sterile conditions, which makes the culture technically demanding.

Growth limitations: The growth and development of organ can be limited by the *in vitro* conditions, as natural (*in vivo*) conditions may not be fully replicated.

Risk of contamination: The culture is susceptible to microbial contamination, which can compromise the efficiency of the process.

Limited scalability: The process has challenge of scalability compared to callus culture, which limits its use for large-scale applications.

3. Embryo Culture

Embryo culture involves isolation and growth of embryos under controlled and artificial environment. This technique is particularly useful for rescuing embryo from seed development that might otherwise fails to develop due to hybridization of incompatible plant types. The step-wise procedure of embryo culture includes the following phases.

Selection of explant: An immature or mature embryo is rescued and used for further *in vitro* growth.

Sterilization of explant: The explant (seeds containing the embryo) is surface sterilized to eliminate contaminants. Typically, 70% ethanol followed by sodium hypochlorite (0.1% for 10 minutes) treatments is used. The explant is rinsed with sterile distilled water to remove any residual agents.

Isolation of embryo: Embryo is rescued under sterile conditions by dissecting seed to carefully extract the embryo. This requires precision to avoid any damage the delicate embryo.

Preparation of culture medium: Sterile nutrient medium (e.g., Murashige and Skoog medium) supplemented with necessary plant growth regulators as well as other additives like vitamins, amino acids, and carbohydrate is prepared. For example, wheat embryo culture requires a medium with a balanced ratio of auxin and cytokinin along with sucrose as a carbon source.

Inoculation: Isolated/rescued embryo is placed on the medium in sterile culture vessels.

Incubation: The culture is incubated under controlled environmental, typically at 25 °C with a 16 hour photoperiod (light and dark cycle).

Embryo growth and development: The rescued embryo growth is monitored for normal seedling development. The embryo germinates and develop into plant under optimal conditions.

Acclimatization of regenerated plant: Once the plantlets achieve sufficient development, they are transferred to soil and gradually acclimatized to external environmental conditions before being planted in the field.

Advantages of Embryo Culture

Hybrid rescue: Allows the rescue of embryo from hybrid crosses that may not develop naturally, enabling the creation of novel hybrids.

Overcoming seed dormancy: It allows bypassing seed dormancy, facilitating the germination of seed that is otherwise difficult to propagate.

Genetic conservation: Useful for conserving the genetic material of endangered or rare plant species by the propagation of embryo that might not survive otherwise in nature.

Research applications: Provides opportunities for studying embryonic development, gene expression, and responses to various treatments.

Uniformity in growth: Ensures uniform growth and development of plantlets, which might be advantageous for experimental consistency.

Disadvantages of Embryo Culture

Technical expertise: it requires precision and expertise in isolating and handling of embryos, which makes the process technically demanding.

Risk of contamination: the culture is susceptible to microbial contamination, which might result in loss of embryos/cultures.

Resource intensive: it involves costly setup of sterile conditions, specialized equipment, and growth media for successful cultures.

Labour-intensive: Requires careful monitoring and continuous maintenance of culture, making the culture labour-intensive and costlier.

Species-specific protocols: Protocol for every species need to be standardized, as the procedure/requirements may vary significantly for different plants.

4. Cell Suspension Culture

Cell suspension culture (growing dispersed plant cells or cell aggregates in liquid nutrient medium) is used for large-scale production of secondary metabolites, studying cell biology, genetic modification, and producing genetically uniform plants. The step-wise process of suspension culture includes the following phases

Selection of explant: To start the culture, callus is induced on the explant (leaf, stem, or root) using solid nutrient medium supplemented with necessary plant growth regulators (like 2,4-D and kinetin).

Establishment of suspension culture: To initiate suspension culture, a small piece of actively growing callus is put into liquid MS medium containing the same growth regulators in a culture flask on a rotary shaker set at 100–120 revolution per minute (rpm) for even suspension culture.

Incubation of culture: The culture is incubated under controlled conditions, typically at 25 °C with 16 hour photoperiod (light and dark cycle).

Subculturing: When sufficient cellular growth occurs in the culture, it needs subculture by transferring a portion of the suspension culture to fresh medium every 7–14 days to maintain cell viability and promote continuous growth.

Cell proliferation: Proliferation of cells in the culture is monitored for a fine suspension of single cell and small cell aggregates.

Harvesting of cells: The fast-proliferating cells, typically at the exponential phase of the growth, are harvested/collected for downstream processing (production of secondary metabolites). For plant regeneration, the cells or cell aggregates are transferred to solid medium with appropriate growth regulators for shoot and root organogenesis.

Advantages of Cell Suspension Culture

Mass proliferation: To enable large-scale continuous production of cells and secondary metabolite production, this is a desired technique.

Uniformity of cells: It provides a homogeneous and uniform population of cells, which is needed for experimental consistency and reproducibility.

Rapid growth: Cells in suspension culture typically grow faster than those on solid media, leading to quicker/faster biomass production.

Disadvantages of Cell Suspension Culture

Technical expertise: It requires technical skill to establish and maintain the culture, including aseptic conditions and growth.

Risk of contamination: The culture is susceptible to microbial contamination, which may compromise the benefits of the culture.

Resource intensive: It involves higher cost due to the need of sterile conditions, specialized equipment, and continuous agitation.

Cell viability: Maintaining cell viability and preventing cell death over extended periods can be challenging tasks.

Limited to certain species: Not all the plant species or cell types are amenable to successful suspension culture, requiring species-specific optimization.

5. Protoplast Culture

Protoplast culture involves isolation and culture of plant cells from which cell walls have been enzymatically removed. The protoplast can be regenerated with cell wall and grown to develop into whole plant. This technique is needed for genetic manipulation, somatic hybridization, and studying cell biology at cellular level. The steps in protoplast culture include the following processes.

Selection of explant: Healthy, young leaf tissue is used to isolate cells for preparing protoplasts.

Sterilization of explant: The explant is surface sterilized by immersing in 70% ethanol followed by treatment with 1% sodium hypochlorite for 10 minutes. The explant is rinsed with sterile distilled water to remove traces of sterilizing agents.

Isolation of protoplast: The sterilized explant is cut into small strips (1–2 mm).

The strips are incubated in enzymatic solution containing cellulase and pectinase in an osmotic stabilizer like mannitol or sorbitol. The enzyme solution digests cell wall, releasing the protoplast. The process is incubated with gentle shaking for 4–6 hours at 25–28 °C.

Purification of protoplasts: Filter the enzyme mixture using a sterile nylon mesh to remove undigested tissue. Centrifuge the filtrate at low speed (100–200 g) to pellet down the protoplasts. The protoplasts are washed by resuspending them in a washing (mannitol) solution and centrifugation.

Viability testing: Viability of the isolated protoplasts is tested using staining with fluorescein diacetate (FDA) stain. Only viable protoplasts appear fluorescent under UV light.

Culturing of protoplasts: Viable protoplasts are resuspended in suitable culture medium, (modified MS medium supplemented with osmotic stabilizer and necessary plant growth regulators like auxins and cytokinins). The culture is incubated in dark under controlled environment at 25 °C to encourage cell wall regeneration and cell division.

Regeneration of cell: Cell wall is regenerated on protoplast and starts dividing to form microcolonies. As the cells grow, osmotic concentration of the medium is reduced gradually. Eventually, the regenerated cells are transferred to a medium to promote shoot and root formation.

Acclimatization of regenerated plants: Once plantlets are formed, they are transferred to soil or a suitable substrate and acclimatize gradually to greenhouse conditions before moving them to the fields.

Advantages of Protoplast Culture

Genetic engineering: it helps facilitating direct insertion of DNA into cells for genetic transformation.

Somatic hybridization: it allows fusion of protoplasts from different species or varieties to create somatic hybrids, which enable combining desirable traits from both the parents.

Cell biology studies: Enables detailed studies on cell biology, including membrane biology, cell wall regeneration, and cellular responses to various treatments.

Disadvantages of Protoplast Culture

Technical expertise: It requires technical skill and expertise on enzymatic treatment, sterilization, and handling of protoplasts.

Risk of contamination: It involves higher risks of contamination, which compromise viability and regeneration potential of protoplasts.

Osmotic stress: Protoplasts are highly sensitive to osmotic stress, and maintaining the correct osmotic pressure is critical for their survival and growth.

Labour-intensive: The process is labour-intensive, requiring careful monitoring and maintenance.

Species-specific protocols: Different plant species and tissues require specific protocols for protoplast isolation and culture, necessitating optimization of protocols.

6. Anther/Pollen Culture

Anther or pollen culture involves *in vitro* propagation of pollen grains to produce haploid plant, which is then doubled to create homozygous line. The steps involved in anther/pollen culture are as mentioned below.

Selection of donor plant: Healthy, vigorous plants at reproductive stage with anthers containing microspores at uninucleate to early binucleate stage are selected to collect explants.

Collection of explat: Panicles are harvested from the donor plants and surface sterilize by immersing them in 70% ethanol for a few seconds followed by a treatment with 1% sodium hypochlorite for 10–15 minutes. The explants are rinse thoroughly with sterile distilled water to remove traces of sterilizing agents.

Isolation of anthers: Under sterile conditions, spikelets are dissected at the end to remove anthers using fine forceps while avoiding any damage to anthers during extraction (Figure 13.4).

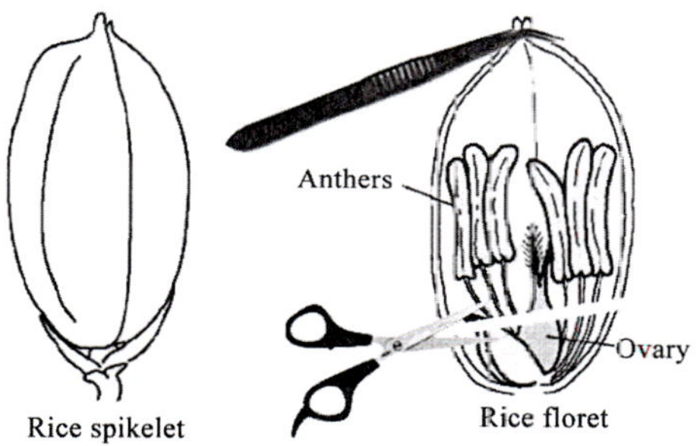

Fig. 13.4: Isolation of anthers from rice spikelet.

Pre-treatment

Some of the species may require a pre-treatment, such as cold treatment (4 °C for 1–7 days), to enhance induction of androgenesis (development of haploid plants from anthers).

Inoculation on anther culture medium: Isolated anthers are placed on a anther culture callus induction medium. For rice, a commonly used anther culture medium is N6 basal medium supplemented with plant growth regulators like 2,4-D (auxin) and kinetin (cytokinin). The culture is incubated in dark at 25–28 °C to promote callus formation or direct embryogenesis from the microspores inside the anthers.

Plantlet regeneration: Callus is transferred to differentiation medium (containing auxin and cytokinin) to promote shoot and root organogenesis. Plants are acclimatized gradually to *ex vitro* conditions before transferring them to soil in fields.

Chromosome doubling: To convert haploid plants to diploids, plantlets are treated with colchicine or any other doubling agent. This produces homozygous lines, which are valuable in breeding programs.

Advantages of Anther/Pollen Culture

Haploid production: This technique can be used to rapidly produce haploid plants, which can be doubled to create homozygous lines; thus, speeding up the breeding process.

Genetic uniformity: It produces genetically uniform lines, which are valuable for genetic studies and breeding programs.

Disadvantages of Anther/Pollen Culture

Species-specific protocols: Anther culture requires specific protocols for different species and even different varieties, necessitating optimization for each case.

Lower efficiency: The efficiency of haploid induction and regeneration can be low and not all the anthers/pollen will respond to the culture conditions.

Technical expertise: It requires technical skill and expertise in handling and culturing anthers/pollen grains.

Risk of contamination: Like other tissue culture techniques, anther/pollen culture is also susceptible to microbial contamination, which further reduces efficiency of the technique.

Applications of Tissue Culture

1. ***Plant Breeding and Crop Improvement:*** Tissue culture techniques enable rapid propagation of plants with desirable traits. Somaclonal variations may also introduce new genetic variations useful for breeding programs.

2. ***Clonal Propagation:*** Micropropagation allows the production of a large number of genetically identical plants from a single explant. This is particularly useful for maintaining uniformity and quality of horticultural crops.

3. ***Production of Virus-Free Plants:*** Meristem culture is used to produce virus-free plants, as meristems are often free of viral infection. This technique can be utilized for vegetative propagation of healthy plants of crops like bananas and orchids.

4. ***Conservation of Rare and Endangered Species:*** Tissue culture techniques like cryopreservation and *in vitro* storage can help conserve genetic material of rare, endangered, and threatened plant species by maintaining them in a viable state for long period.

5. ***Genetic Engineering:*** Tissue culture is a prerequisite for genetic engineering and transformation processes. Protoplast culture and somatic embryogenesis facilitate the introduction of new genes into plant cells, enabling the development of genetically modified crops.

6. ***Production of Secondary Metabolites:*** Cell suspension cultures and hairy root cultures are employed to produce valuable secondary metabolites like pharmaceuticals, flavors, fragrances, and dyes, in a controlled environment.

7. ***Somatic Hybridization:*** Protoplast fusion techniques allow the creation of somatic hybrids by fusing protoplasts from different species or varieties. This can combine desirable traits from both parents, creating novel hybrid plants that are difficult/impossible to produce through conventional breeding.

8. ***Embryo Rescue:*** Embryo culture is used to rescue hybrid embryo from incompatible crosses that might otherwise aborts. This technique is instrumental in plant breeding to create hybrids between species that do not naturally cross.

9. ***Haploid Production and Doubled Haploids:*** Anther and pollen culture techniques are used to produce haploid plants, which can be doubled to create homozygous lines. This accelerates the breeding process and helps developing pure lines for research and commercial purposes.

10. ***Somatic Embryogenesis:*** Somatic embryos can be produced from somatic or non-reproductive cells using tissue culture techniques. Somatic embryogenesis is valuable for cloning plants, studying embryogenesis, and producing synthetic seeds.

11. ***Production of Synthetic Seeds:*** Synthetic seeds are made of encapsulated somatic embryos, which can be used for germplasm conservation, exchange, and an alternative to traditional seed propagation methods.

12. ***Commercial Horticulture:*** Tissue culture techniques are widely used in commercial propagate of horticulture and ornamental plants, flowers, and high-value crops, ensuring uniformity, disease-free plants, and rapid production cycles.

Let's Check Your Understanding

1. Which tissue culture technique is primarily used to produce virus-free plants?

    a. Somatic embryogenesis
    b. Meristem culture
    c. Protoplast fusion
    d. Anther culture

2. What is the main advantage of using anther or pollen culture in plant breeding?

    a. Production of virus-free plants
    b. Conservation of endangered species
    c. Rapid production of haploid and homozygous lines
    d. Large-scale production of secondary metabolites

3. Which application of tissue culture is crucial for the commercial propagation of ornamental plants and flowers?

    a. Embryo rescue
    b. Micropropagation
    c. Protoplast fusion
    d. Somatic hybridization

4. In the context of tissue culture, what is the purpose of somatic hybridization?

    a. Producing virus-free plants
    b. Creating genetically identical plants
    c. Combining desirable traits from different species or varieties
    d. Producing synthetic seeds

5. Which tissue culture technique involves the production of synthetic seeds from encapsulated somatic embryos?

    a. Embryo culture
    b. Cell suspension culture
    c. Meristem culture
    d. Somatic embryogenesis

6. Tissue culture techniques enable the rapid propagation of plants with desirable traits, such as __________ resistance, drought tolerance, and improved yield.

7. ________________ is used to produce virus-free plants, crucial for the health of vegetatively propagated crops like bananas and orchids.

8. Protoplast fusion techniques allow the creation of somatic hybrids by fusing protoplasts from different species or varieties, combining desirable traits from both the ________________.

9. Anther or pollen culture is used to produce haploid plants, which can be doubled to create ________________ lines quickly, speeding up the breeding process.

10. Synthetic seeds, made from encapsulated somatic embryos, can be used for germplasm conservation, exchange, and sowing, providing an alternative to traditional ________________ propagation methods.

Answers

1. b
2. c
3. b
4. c
5. d
6. Disease
7. Meristem culture
8. Parents
9. Homozygous
10. Seed

14

Testing Seed Viability

Introduction

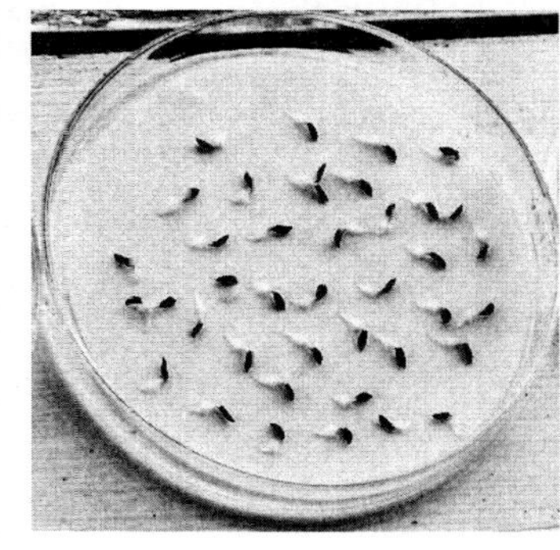

Seed is a living entity (mature ovule) consisting of a tiny plant in embryonic stage along with the stored food protected by an outer layer known as seed coat. Seed is an important biological input in agriculture, which is economically important for successful crop husbandry. Cotyledons serve as the source of nutrients (proteins, carbohydrates, and oils) that not only nurture the plant at the early developmental (seedling) stages of plant but also the vegetarian (heterotroph) organisms. Being one of the important inputs in agriculture, viability of seed needs to be assessed to know the germination percentage of seeds used to raise a crop. Since viability of seeds may deteriorate on storage depending on the storage conditions. Therefore, often it becomes important to test viability/germinability and vigour of seeds before being used for raising a crop or conducting an experiment. In this chapter, we briefly discuss about the different methods which can be used for testing the viability of seeds.

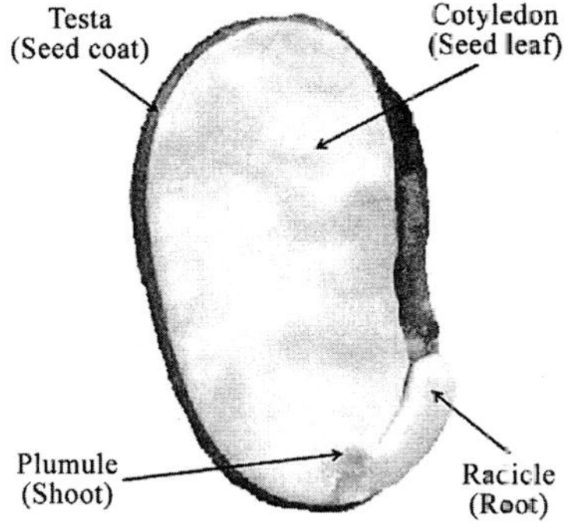

Fig. 14.1: Parts of a seed.

Seed Viability

Seed viability can be defined as the capability of the seed to germinate and produce a normal seedling within a specified period. In other words, seed viability indicates how many seeds in a lot are alive and could develop into plants under appropriate field conditions. It is very important that most of the stored seeds are capable of germinating when sown in a field. They must possess higher viability at the beginning of storage and maintain it over the storage. A seed with high initial viability survives longer in storage; however, the viability deteriorates slowly at first and then rapidly as the storage prolongs. It is important to know the duration of storage after which a considerable decline in viability occurs under given storage conditions so that to take necessary action to revive the seeds. Excessive deterioration in vianility of seeds may lead to loss of the seed/material/ germplasm.

Factors Affecting Seed Viability

There are two types of factors that affect viability of seeds.

(i) Pre-harvest factors,

(ii) Post-harvest factors.

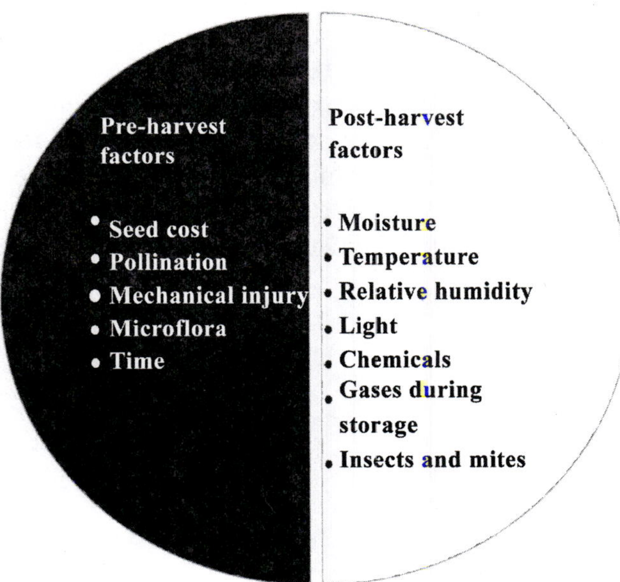

Fig. 14.2: Factors affecting viability of seeds.

Tests to Determine Seed Viability

Several different types of tests (dissecting, germination, moisture content, embryo excision, radiography, using tetrazolium, hydrogen peroxide, etc.) are deployed to determine the viability of seed. However, among these tests, three tests are commonly used for seed viability assessment.

(i) Tetrazolium test,

(ii) Germination test,

(iii) Excised embryo test.

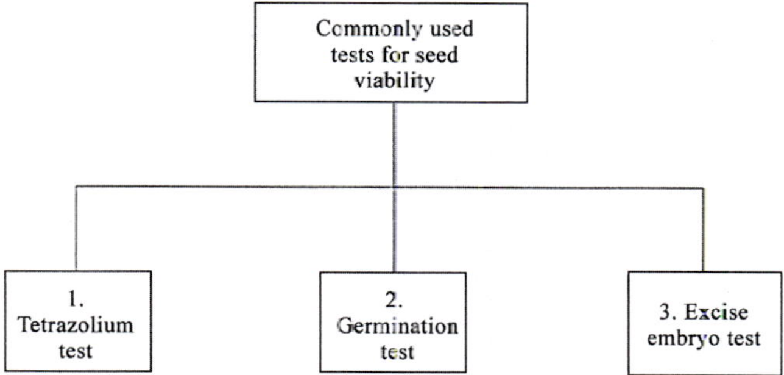

Fig. 14.3: Commonly used tests for viability of seed.

1. Tetrazolium test: Tetrazolium test is the most commonly used rapid and accurate method for testing seed viability. It was developed by Lakon, wherein tetrazolium salt is used to prepare a colourless 0.1−1.0% solution having the pH ~7. This test can be used to determine whether a dormant seed is viable or not (that can't be determined using germination test).

Principle: All living tissues respire and are capable of reducing the colourless 2,3,5-triphenyl tetrazolium chloride into a red coloured formazon compound. This reaction is catalyzed by dehydrogenase enzyme produced during respiration in live tissue/seed. The formation of red colour indicates the seed is alive. Though tetrazolium enters into both living and dead cells but only living cells (producing dehydrogenase enzyme) catalyze the formation of formazan, which stains the viable seeds red whereas the absence of the enzyme/respiration prevents the production formazan leaving the dead seed (aged tissue) remain unstained.

Evaluation of Monocot Seeds

If embryo and scutellum are completely stained (or extreme tip of scutellum alone unstained), then seed is categorized as viable/germinable seed. If the embryonic axis is completely unstained (or part of the scutellum and embryo are unstained), then the seed is categorized as a dead/non-germinable seed.

Procedure of Tetrazolium Test

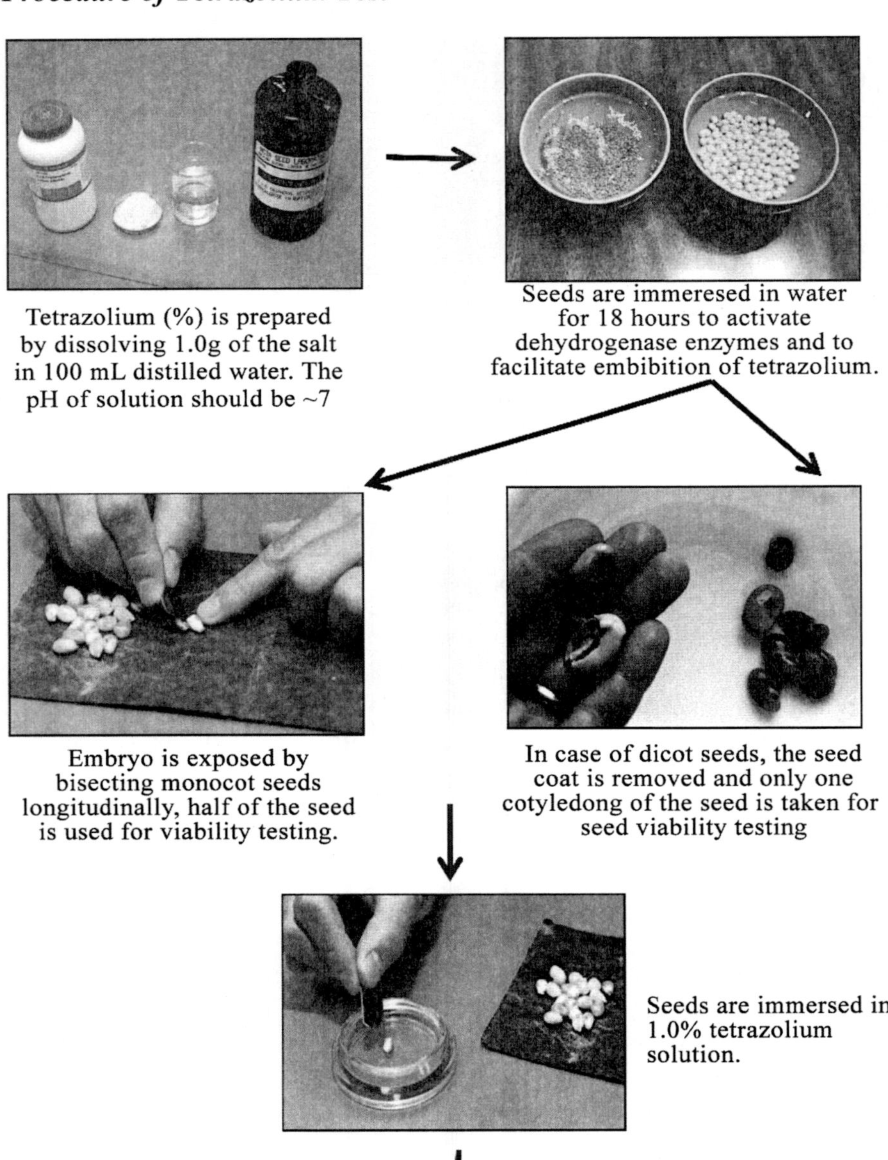

Tetrazolium (%) is prepared by dissolving 1.0g of the salt in 100 mL distilled water. The pH of solution should be ~7

Seeds are immeresed in water for 18 hours to activate dehydrogenase enzymes and to facilitate embibition of tetrazolium.

Embryo is exposed by bisecting monocot seeds longitudinally, half of the seed is used for viability testing.

In case of dicot seeds, the seed coat is removed and only one cotyledong of the seed is taken for seed viability testing

Seeds are immersed in 1.0% tetrazolium solution.

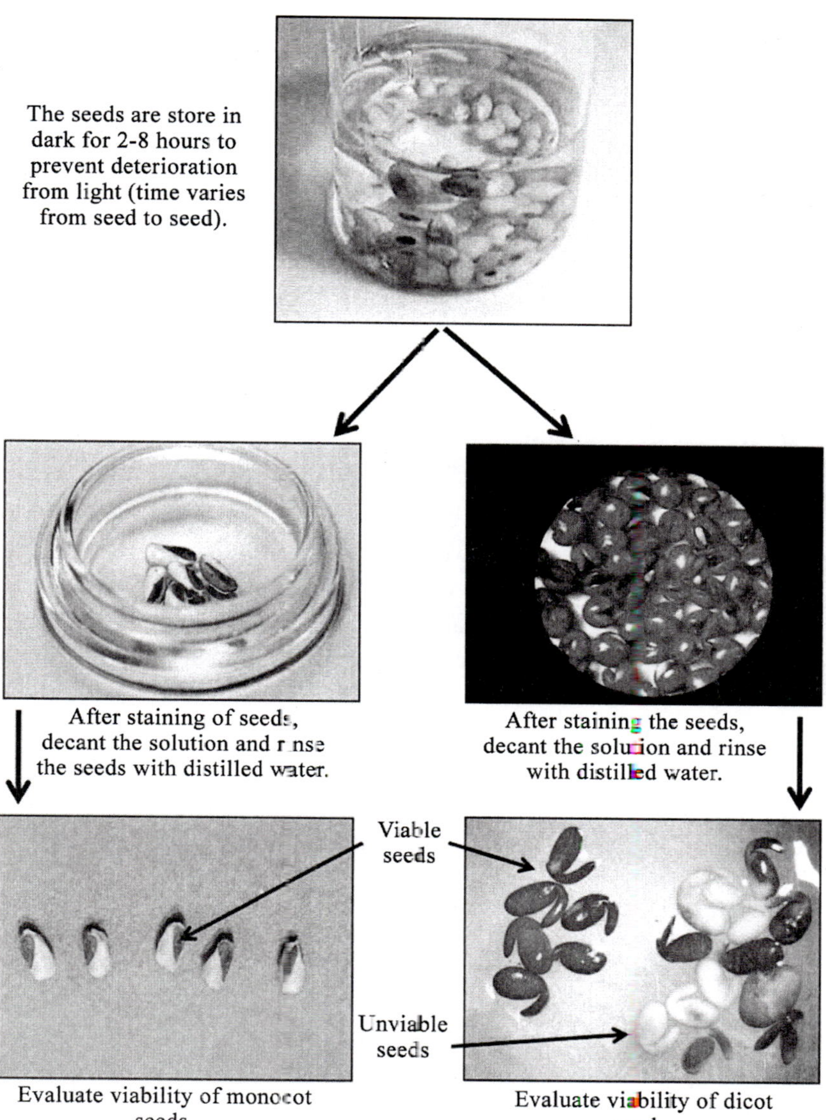

Fig. 14.4: Procedure of tetrazolium test of seed viability.

Evaluation of Dicot Seeds

If a seed is completely stained (or the extreme tip of the radicle alone is unstained), then the seed is categorized as a viable/germinable seed. If more than the extreme tip of the radicle is unstained (or more than half of the cotyledon area is unstained or the entire seed is completely unstained), then the seed is categorized as a dead/non-germinable seed.

Calculation of Seed Viability

Seeds that get stained (viable/germinable seeds) are counted and used for the calculation of seed viability percentage using the following formula.

$$Seed\ viability\,(\%) = \frac{\text{No. of seeds taken stain}}{\text{Total seeds taken for staining}} \times 100$$

Precautions During Tetrazolium Test

The pH of tetrazolium solution should be ~7. Solution having pH >8 or <4 might result in either intense staining or may not stain even the viable seeds. If the water taken to dissolve tetrazolium salt is not neutral, then use phosphate buffer with pH 7.0. Care should be taken while bisecting monocot seeds and removing the seed coat of dicot seeds to avoid any damage to embryonic axis.

***2. Germination test*:** This test is performed to determine what proportion of seeds will germinate under favourable conditions to produce normal seedlings (having the essential structures like roots and shoot) capable of developing into plants. It is the most accurate/reliable method of testing viability of seeds. About 200 seeds are taken for the viability/germination test at the beginning of storage. If the test results show germination to be less than 90%, additional 100 seeds are taken in 2 replications to confirm the test result/method. The seed viability is recorded using the mean of two tests. Different methods of germination test are as follows:

1. Top-of-paper method,
2. Between-paper method,
3. Germination on sand.

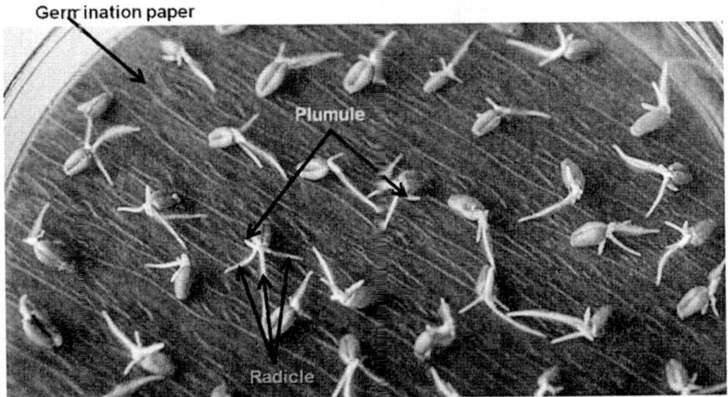

Fig. 14.5: Top-of-paper germination test (wheat).

Fig. 14.6: Between-paper method of germination test.

Fig. 14.7: Germination on sand method.

Procedure: The seeds are spread evenly on the surface of wet medium (germination paper or sand) in such a way that seeds are away from each other. The medium should contain optimum amount of moisture for the seeds to germinate. The setup is then kept in a germination chamber to provide optimum conditions of light, temperature, and humidity for the seeds to germinate. The test is continued until (recommended time for germination of the seeds) viable seeds get germinated.

Calculation

Viability (%) = (G + F + A) ÷ X × 100

Where:

G = number of seeds germinated

X = number of total seeds sown taken for the test

F = number of fresh ungerminated seed

A = number of abnormal seedlings

Precautions

The medium should be disinfected to prevent any fungal infection during the germination test, otherwise it may lead to the failure of the test. The temperature and relative humidity of the germination chamber/room should be maintained as per the seed/species.

3. Excised Embryo Test: The excised embryo test is similar to germination tests in that it measures the quality of the seed by their actual germination and it allows some measure of the embryo dormancy to be made, by counting those seeds which, although not growing normally, have grown slightly, remained firm and have kept their colour for the test period. The excised embryo test is quick, seven to ten days, and accurate.

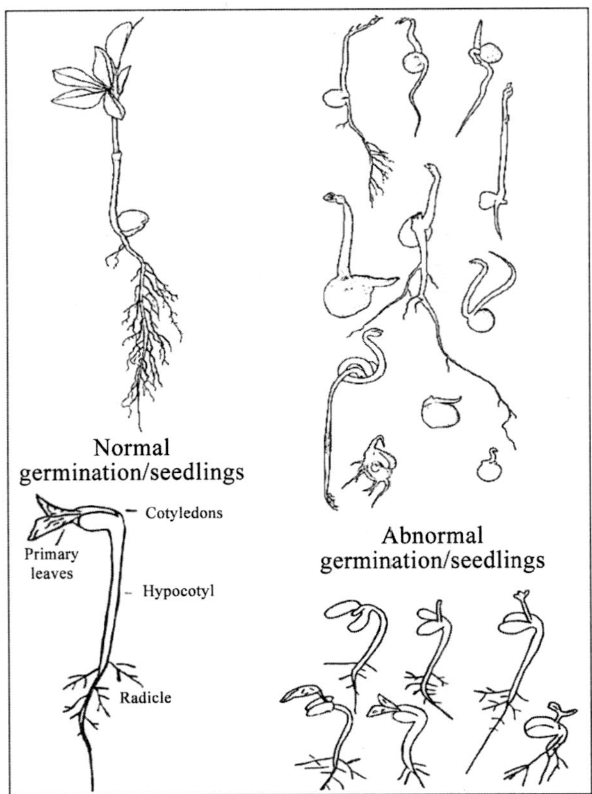

Fig. 14.8: Normal and abnormal germination/ seedlings. The normal seedling has all essential structures (roots, shoot, and other parts) capable of developing into normal plant, abnormal seedlings lack one or more essential structures which may lead to death of the seedling.

Principle: This method is commonly used to test the viability/germination of seeds of woody species whose embryos require a period of dormancy/after-ripening before it can germinate. Since the seed coat is impermeable to water or the embryo requires a after-ripening period, the embryo must be excised to test viability.

Procedure

The test is performed on 200 seeds from a lot. Following steps are taken:

1. Surface sterilize the seeds.
2. Cut 1 mm off at the radicle end of the seed.
3. Soak the seed in water for 18 hours.
4. Cut the seed on both sides of the embryo.
5. Try opening the seed to expose the embryo.
6. Remove the embryo with the help of a needle.
7. Incubate the embryo for nine days at 20 °C.

Live/viable embryo will spread open the cotyledons and they will become green. The dead embryo will deteriorate gradually.

Precautions: The exposed embryo is susceptible to microbial attack; therefore, good sanitary conditions should be maintained during the testing process. Seeds are considered to be viable if the cotyledons turn green and spread apart.

Limitations: This test is not valid for previously germinated seeds and must not be applied to the samples that contain any dry germinated seeds. The success of this test requires considerable skill and experience of the researcher/tester. As per the ISTA rules, it is restricted to only a certain species. In a study by Schubert (1965), a comparison between the excised embryo method and the tetrazolium method of determining viability of dormant tree seeds was performed. The study concluded that tetrazolium method should be preferred, over the embryo excision method, with improvement in the tetrazolium test by using bactericides and stronger reducing solution to enhance staining of tissues.

Conclusion

Seed viability test is an analytical method to evaluate the potential of seed for germination. Seed viability is not only associated with the success of the crop but also with the input cost and yield. Hence, this has received considerable attention to maximize yield and profits from crop husbandry. However, different methods mentioned in this chapter have some benefit as well as limitations as some of the test is ineffectiveness on dormant seeds. This issue can be resolved using tetrazolium test, which is based on seed viability regardless of the dormancy status of the seed.

Let's Check Your Understanding

1. Seed germination test for viability can be performed by
   a. Top-of-paper
   b. Between-paper
   c. Germination on sand
   d. All of the above.
2. Seed is the only biological input in agriculture.
   True / False
3. A seed cut into two halves and immersed in 0.1% triphenyl tetrazolium chloride solution. Its viability is indicated by the development of colour as
   a. Yellow
   b. Red
   c. Blue
   d. Green
4. How will you calculate seed viability percentage after performing tetrazolium test?
5. How will you calculate seed viability percentage after performing germination test?
6. Tetrazolium is a colourless solution that is used at the concentration of .... with a pH of around ....
7. The formation of red colour in tetrazolium test indicates the seed is alive.
   True / False
8. In case of monocot seeds, if the embryonic axis is completely unstained or part of the scutellum and embryo are unstained, then the seed is categorized as a .... seed.
9. What are the factors affecting seed viability?
10. Seed coat is also known as .. ..?

Answer

1. d
2. False
3. b
4. $Seed\ viability(\%) = \frac{\text{No. of seeds taken stain}}{\text{Total seeds taken for staining}} \times 100$
5. Viability (%) = (G + F + A) ÷ X × 100
6. 0.1−1.0%, 7
7. True
8. unviable or non-germinable
9. (i) Pre-harvest factors

   (ii) Post-harvest factors
10. Testa

15

Testing of Pollen Viability

Introduction

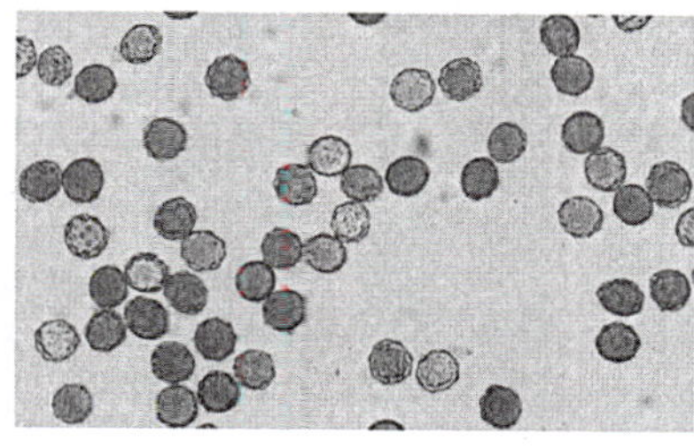

Test of pollen viability (extent of pollen viability) is important in pollination experiments to ensure fertility/seed setting. It is also important in monitoring state of pollens during storage in ecological/ taxonomic studies as well as during research on pollen biochemistry, genetics, stigma interactions, and incompatibility studies. Pollen viability refers to the ability of pollen to accomplish post-pollination events to achieve fertilization. Pollen viability test has gained importance particularly in plant breeding and hybrid seed production experiments/ practices. In this chapter, you will be acquainted with the basics of pollen viability test. However, for more professional knowledge on the topic readers are suggested to refer specialized books.

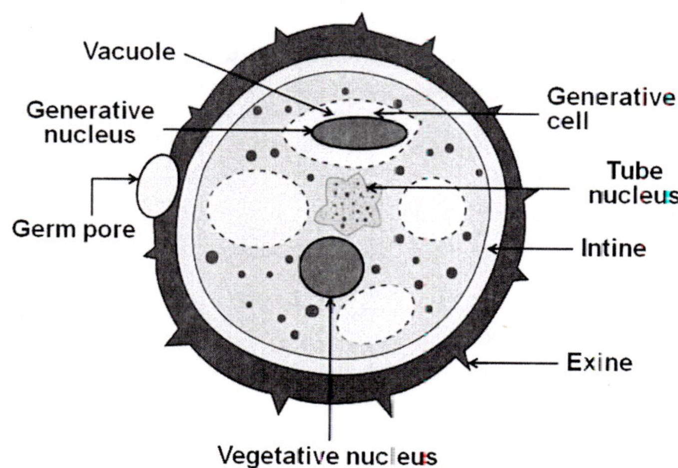

Fig. 15.1: Parts of a pollen.

Pollen Viability Test

It is important to know the extent of viability of pollens used in pollination experiment, in finding fertility of a plant of interest, to monitor state of pollen on storage, in ecological/taxonomic studies, as well as in research on biochemistry, genetics, stigma interaction, incompatibility assay, etc. of pollen. Several pollen viability tests have been developed time-to-time. It is recommended to initially use different tests for a given pollen type to find out the most appropriate method that provides reliable result on pollen viability. It is important to note that every procedure requires standardization of protocol according to the sample used. Following different tests are used for assessing pollen viability.

1. In-vitro Pollen Germination Test

It is a simple, speedy, and quantitative test for assessing pollen viability. In this method, a collection/sample of pollen is germinated *in vitro* and observed under microscope. The percentage of germinating pollens (producing pollen tube) after appropriate treatment and a given period of time is determined. The percentage provides an index of viability of the pollen sample.

Procedure

(i) A germination medium is prepared using optimal concentration of sucrose (depending on the species), boric acid, and calcium nitrate. Standard pollen culture medium (e.g., Brewbaker and Kwack medium, Roberts' medium, Hodgkin and Lyon's Medium, PEG medium, EACA medium, Raffinose medium, Agarose medium, etc.) can also be used. The pH of medium and the incubation temperature are some of the important factors affecting pollen germination.

(ii) Fresh pollen grains are collected a day after anthesis, scattered uniformly on a drop of medium in a Petri dish, and covered with the lid lined with moist filter paper. The plate is incubated in a BOD incubator at temperature (18 °C – 25 °C) suitable for the pollen/plant species.

(iii) After incubation, a drop of stain (e.g., Calberla's solution, acetocarmine, Sudan IV, Alexander's solution+ lactic acid) is added to the droplet and the droplet is transferred to a microscopic slide for scoring/counting under microscope.

(iv) The total number of pollen grains and the number of germinated pollens are recorded. At least 200 to 300 pollen grains from 5 to 10 microscopic fields should be scored for each treatment. The length of pollen tube can be measured using an ocular micrometer. Microscopic illumination should be adjusted for clear visibility of pollens and pollen tubes.

(v) Viability/response of the cultured pollen grains is assessed in terms of percent pollen germination and average pollen tube length. A pollen grain is considered germinated when the length of its tube is greater than the diameter of the pollen grain.

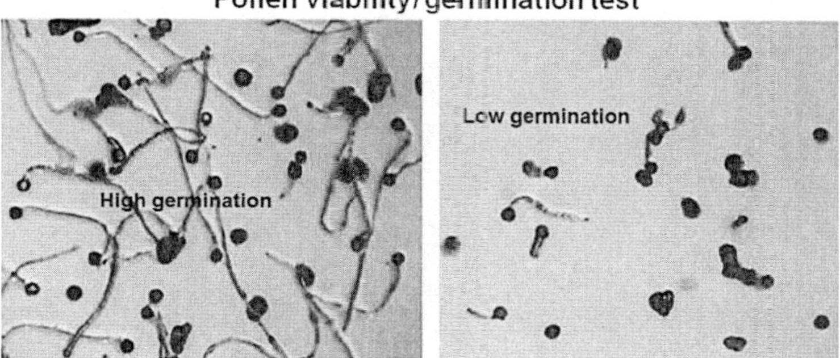

Fig. 15.2: Pollen viability/germination test.

Limitations

i. In some species, especially in three-celled pollen systems like wheat, higher/satisfactory germination percentage is less likely to be recorded. *In vitro* growth conditions do not always simulate with the *in vivo* conditions. Also, the medium which stimulates optimal germination of fresh pollen may not be suitable for the stored pollen.

ii. Pollen grains of many species exhibit population effect; therefore, pollen density should be kept optimal during the test. Smaller population of pollen grains might not show satisfactory germination due to insufficiency of pollen growth factors. Very large number of pollens may also give unsatisfactory results, as nutrients in the culture medium may become limiting factor.

iii. Observations taken on the pollen not uniformly scattered/distributed in the medium droplet and/or using inadequate number of replicates are likely to give erroneous results.

2. Pollen Germination on Stigma

In case of controlled compatible pollination, germination of pollen on the stigma and growth of pollen tube in style are used for pollen viability test. The amount of pollen tubes growing in the style indicates the efficacy of the pollen that influence fruit set and seed set. In case of *Brassica oleracea*, pollens are considered fully viable if at least 70% pollens are found growing in the style. All pollen grains do not germinate, the percentage of pollen grain germination and tube growth depends on external and internal factors, External factors include temperature, light, humidity, while internal factors include availability of carbohydrate, boron, calcium, enzymes, hormones, etc.

Materials required: An opened matured flower, Coverslips, Glass slide, Needles, Safranin stain, Glycerine, and Microscope.

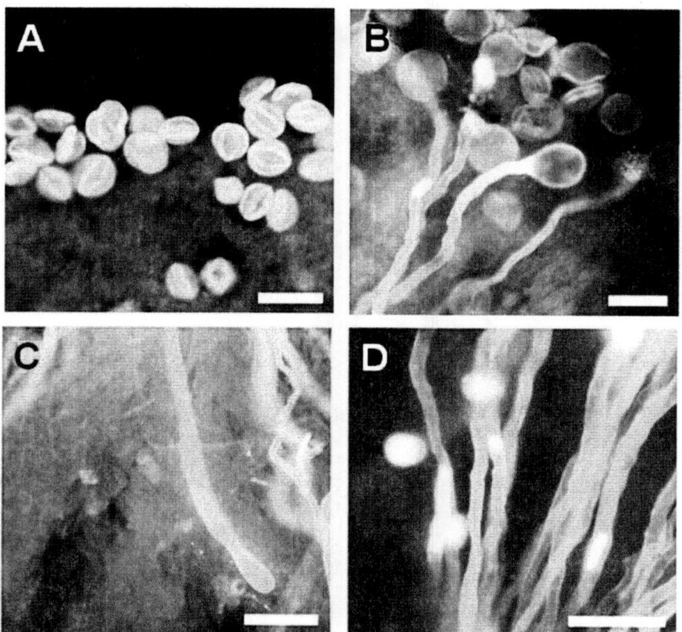

Fig. 15.3: Pollen germination on stigma. **(A)** Pollen adhesion on stigma, **(B)** pollen germination on stigma, **(C)** Pollen tube arrested in the style due to incompatibility reaction, **(D)** Pollen tubes advancing to the base of style.

Procedure

1. Separate pistil from the flower, take the stigma and place it on a clean slide,
2. Stain the stigma with safranin and let it sit for 5 minutes, remove any excessive stain using blotting paper,
3. Mount the stained stigma in glycerine and place a cover slip over it using forceps and needle,

4. Gently press the cover slip/stigma, and remove excessive stain using blotting paper,
5. Observe the slide under a microscope. Review it every 30 minutes to record the observations.

Observations

1. Many pollen grains can be seen on stigma.
2. A thin pollen tube protruding out through germ pore can be seen.
3. The pollen grains that do not develop with pollen tube are considered as non-viable.

3. Fruit Set and Seed set test

This is the most reliable way of testing pollen viability and it is expressed in terms of the ability of pollen to effect fertilization resulting in seed and fruit set. Pollen viability is the ability of pollen to effect fertilization and development of fruit/seed; hence, considered to be the most authentic test for pollen viability. In this test, seed set is determined following the controlled pollinations of stigma compatible with the test pollen sample. This method also provides a measure of compatibility between pollen and stigma, indirectly through plotting saturation graph.

Limitations

× The procedure is laborious and time-consuming (as it takes weeks for fruit and seed set).

× Moreover, the test can be performed only during the flowering season for the species.

4. Fluorochromatic reaction or Fluorescein di-acetate test

Fluorescein diacetate (FDA) is a non-polar, non-fluorescent solution which readily enters cytoplasm of pollen grains when they are mounted on it. Esterase present in the cytoplasm of pollen grain hydrolyses FDA and releases fluorescein, which is polar and fluorescent compound. In contrast to FDA, fluorescein is polar compound which passes (in small amount) through intact membrane and accumulates in the cytoplasm of viable pollen grains. This gives fluorescent bright green or yellowish green colour on observation under fluorescent microscope. If the cell membrane of vegetative cell is not intact, fluorescein easily comes out into the mounting medium and the pollen grain fails to be seen bright fluorescent. If the pollen grain lacks esterase (which hydrolyses FDA), then also it is not seen fluorescent. Thus, FDA/FCR test assesses two properties of the pollen grains: (i) integrity of the cell membrane of the pollen and (ii) esterase activity, which hydrolyses the fluorescein ester.

Procedure

1. Add a few drops of FDA stock solution to 2–5 mL of sucrose solution in a small glass vial/tube until the resulting mixture shows persistent turbidity.
2. Put a drop of Sucrose-FDA mixture on a micro slide.
3. Suspend a small amount of pollen grains in the drop and ensure that they distribute uniformly in the preparation.
4. Incubate the preparation in a humid chamber (>90% RH) for 5–10 minutes.
5. After the incubation, place in a cover glass and observe the preparation under fluorescent microscope using suitable filters. The pollen grains appear bright fluorescent are scored as viable.

Precautions

The Sucrose-FDA mixture should be used within 30 minutes from its preparation, otherwise most of the FDA would precipitates. FDA is toxic; hence, it should be handled and disposed carefully.

Advantages

1. This test is faster and easier compared to *in vitro* germination test. It gives satisfactory result for pollens from a range of species.
2. It can be used to assess viability of pollen as well intactness of cell membrane.

5. Tetrazolium test

This is the most commonly used test for pollen viability. The tetrazolium salts used for the test include 2,3,5-triphenyltretrazolium chloride (TTC) and nitro blue tetrazolium. The test involves reduction of colourless soluble tetrazolium salt to a reddish insoluble substance called formazan in presence of dehydrogenases. Tetrazolium test performs satisfactorily in assessing pollen viability for several species. However, it does not correlate positively with *in vitro* germination test. Pollen grains may display gradation in colour development (from very light to deep red), which makes difficult to judge a cut-off point for colour intensity in order to determine pollen viability.

Preparation of solution

Prepare 0.2–0.5% TTC in sucrose solution of suitable concentration to prevent bursting of pollen grains. The concentration of TTC may be up to 1% in some cases like Cherry Laurel pollens (*Prunus laurocerasus* L.) and *Jatropha curcas*. Addition of small amount of triazine derivatives to TTC solution in sucrose acts as an intermediate co-factor which facilitates acceptance of hydrogen ions by tetrazolium salt and consequently improves pollen response.

Precaution TTC solution undergoes photo-oxidation and can be easily reduced by light; hence it is necessary to keep the solution and the pollen grains in dark in a brown bottle. It can be stored in the refrigerator for a couple of weeks.

Procedure

1. Take a drop of TTC solution on a microslide.
2. Suspend a small amount of pollen in the TTC drop and spread it uniformly in the drop with the help of a slim brush.
3. Apply a coverslip.
4. Transfer the preparation to a humidity chamber (>95% RH).
5. Incubate the preparation in dark (keeping the preparation in a closed chamber, e.g., table drawer) under laboratory temperature (30–37 °C for 30-60 minutes (in some extreme cases up to two hours).
6. After the incubation, observe the slide under microscope and score it for percentage of viable pollen grains, i.e., pollen grains that have turned red due to accumulation of formazan. Pollen grains stained orange or bright red colour are to be counted as viable.

Precaution: Score pollen grains only in the central area of the slide, as pollen grains near the margin of the cover glass show variable degrees of coloration due to higher oxygen availability.

Safety Measures while Performing Pollen Viability Test

1. Pollen is known to be severe allergen and may be problematic for workers in a laboratory. Plants with anemophilous (wind dispersed) pollen should be kept in enclosed areas and a dust mask must be used while working with such plants.

2. Spilt pollens should be washed away from working areas and from hands and face. Pollen sensitive people must be particularly careful and avoid repeated contact with pollens.
3. In pollen viability tests, mostly non-hazardous chemicals are used but FDA and DAPI are particularly toxic and should be dispensed in a fume hood.

Let's Check Your Understanding

1. Which is the most difficult stage *in vitro* germination of pollen grain?
   - A. Standardizing the germination medium
   - B. Uniform distribution of pollen grains
   - C. Staining and transferring droplet to microscope slide
   - D. Scoring using ocular micrometer.
2. FDA / FCR test was developed by?
   - A. R.B. Knox
   - B. George Lakon
   - C. Bolat and Pirlak
   - D. J. Heslop-Harrison and Y. Heslop-Harrison
3. Which dye is used to stain the stigma in pollen germination on stigma experiment?
   - A. Methylene blue,
   - B. Crystal violet,
   - C. Safranin
   - D. Ethidium bromide
4. In the presence of light, TTC solution undergoes photo-oxidation and can be easily reduced; so, it is necessary to keep the solution in dark in a brown bottle.

   True / False
5. What is the criterion for considering a pollen grain to be germinated?
6. FDA and DAPI are particularly toxic and should be dispensed in a fume hood.

   True / False
7. The tetrazolium test involves reduction of a colourless soluble tetrazolium salt to a .... coloured insoluble substance called ...., in the presence of dehydrogenases.
8. Write full forms of FCR and FDR.

Answers

1. a
2. d
3. c
4. True
5. Pollen grain is considered germinated when the length of its tube is greater than the diameter of the pollen grain.
6. True
7. reddish and formazan
8. Fluoro chromatic Reaction (FCR) and Fluorescein Di-Acetate (FDA)

16

Description of Flowering Plants

Introduction

Flowering plants (Angiosperms) are identified by their morphological characteristics like flower, fruit and seeds that help their taxonomic characterization into different species, genera, families and orders. The floral characters which are used for taxonomic characterization of plants include the type of inflorescences, perianth structure, floral symmetry, arrangement of floral leaves, type of androecium, stamens, gynoecium, carpel, ovule, etc. The type of fruit and seed also play important roles in taxonomic classification. In this chapter we will briefly discuss about various morphological features of flowering plants (in botanical terms) that are useful in taxonomy.

Nomenclature of Plant

Nomenclature of plant is the formal and scientific naming of plants. It is related to but distinct from taxonomy for grouping and classifying plants and provide botanical name.

The scientific/botanical nomenclature started with Species Plantarum by Carolus Linnaeus in 1753. Botanical nomenclature is governed by the International Code of Nomenclature for algae, fungi, and plants (ICN), which replaces the International Code of Botanical Nomenclature (ICBN).

ICBN and ICN

International Code of Botanical Nomenclature, (ICBN) was adopted by the VIIth International Botanical Congress held in Stockholm, July1950. While the International Code of Nomenclature for algae, fungi, and plants (ICN) was adopted by the XVIIIth International Botanical Congress held in Melbourne, Australia in July 2011.

Taxonomic Character

A taxonomic character is any attribute of a member of a taxon by which it differs from a member of a different taxon.

Taxonomic Characteristics

1. Morphology: Plant growth habit, leaf arrangement and shape, flower and fruit characteristics.
2. Anatomy: Secondary xylem characteristics, epidermis including trichomes and stomata.
3. Palynology: Pollen characteristics including size, shape, aperture and exine sculpture.
4. Cytology: The chromosome number, structure and habit
5. Chemosystematics: Alkaloid, phenolic and amino acids contents. Protein, enzyme and DNA.

Flowering Plants

These are the plants that possess flowers/fruits and form the clade Angiospermae. The term "angiosperm" is derived from Greek word "angeion" (means container or vessel) and sperma (means seed), that refers to those plants that produce their seeds enclosed within a fruit.

Classification of Flowering Plants

Flowering plants are classified on the basis of three characteristics:

1. Longevity
2. Habitat
3. Appearance

While the flowering plants are classified into four types based on the longevity:

a. Ephemerals

b. Annuals

c. Biennials

d. Perennials (Figure 16.1).

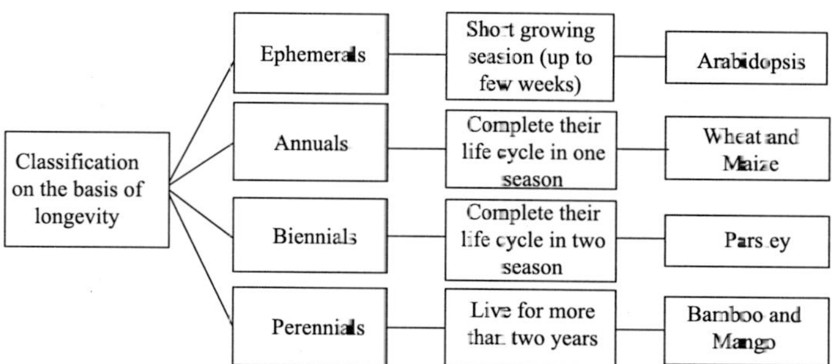

Fig. 16.1: Classification of flowering plants on the basis of longevity.

On the basis of habitat, plants can be of five different types:

a. Hydrophytes
b. Mesophytes
c. Xerophytes
d. Epiphytes
e. Halophytes (Figure 16.2).

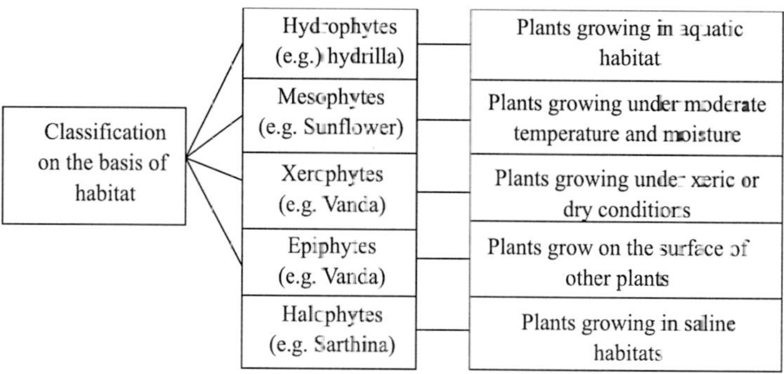

Fig. 16.2: Classification of flowering plants on the basis of habitat.

Similarly, on the basis of appearance, plants can be of three different types:

a. Herbs
b. Shrubs
c. Trees (Figure 16.3)

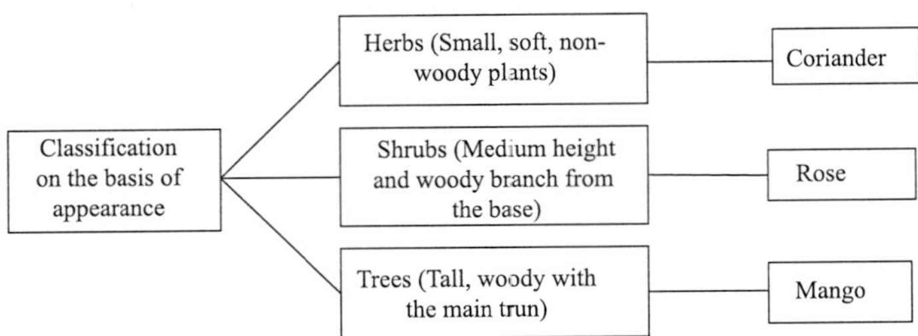

Fig. 16.3: Classification of flowering plants on the basis of appearance.

Morphological Characteristics of Plant

Angiosperms are identified by their morphological characters. They are grouped mainly on the basis of their floral characters. Similarity in the morphology of flowers, fruits and seeds within different species, genera, families, and orders provides a base in characterizing them in taxonomic groups. The floral characters are commonly used in plant taxonomy include the types of inflorescences/flower, perianth structure, floral symmetry, union of floral leaves in each whorl, types of androecium/stamens, gynoecium, carpels, ovules, bracteoles/pedicels, etc. The type of fruits and seed also provide good diagnostic features useful at various levels of taxonomic classification.

A. Leaf

Leaf characters such as arrangement, type, form, duration and venation are widely used in both classification as well as identification of plants (Figure 16.4). In general, leaf can be of two types:

1. Simple leaf
2. Compound leaf

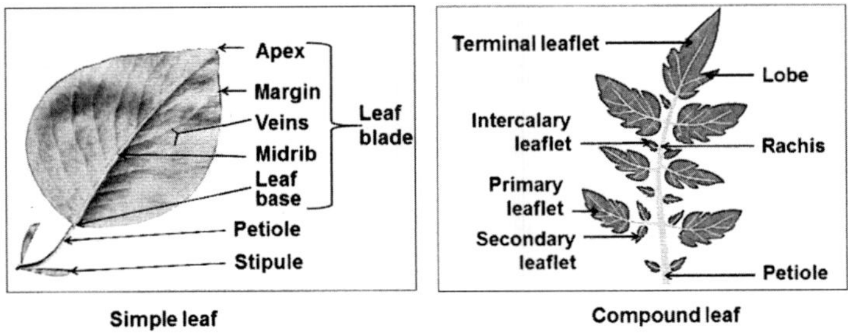

Fig. 16.4: Types of leaf.

Table 1: Difference between simple and compound leaf.

Simple leaf	Compound leaf
Leaf has an undivided leaf blade.	Leaf blade is divided into leaflets.
Leaflets are absent.	Leaflets are present.
Leaf is attached to the stem/branch through petiole	Leaflets are attached to the rachis and petioles.
An axillary bud is present in the axil of a simple leaf.	An axillary bud is present in the axil of a compound leaf but the leaflets of the compound leaf do not have axillary bud.
Example is Mango and Guava leaf.	Example is Rose and Coriander leaf.

B. Flower

Flower comprises of the vegetative and reproductive parts. In addition, there are four whorls that are important/responsible for the arrangement of calyx, corolla, androecium and gynoecium. When a flower possesses both androecium and gynoecium, then it is called as bisexual; and if a flower possesses either stamen or carpels, it is called unisexual.

Petals and sepals are the vegetative parts, while stamen are pistil are the reproductive parts of the flower (Figure 16.5).

Sepal: It is the part of flower that is generally green in colour and protects the growing buds. It is found beneath the petals.

Petals: These are coloured part of the flower, which attracts other organisms. The petals are of different colours and patterns. These provide different looks to the flower.

Stamen: It is also known as androecium and is the male reproductive part of the flower. It is comprises of anthers and filaments. The anthers are the sac-like structure that produces pollens.

Pistil: This is the female reproductive part of a flower, also known as gynoecium. It is comprises of stigma, style and ovary. Stigma is the tip of the carpel, style is a stalk that connects stigma and ovary, and ovary is an enlarged basal part of the pistil/flower which holds ovules.

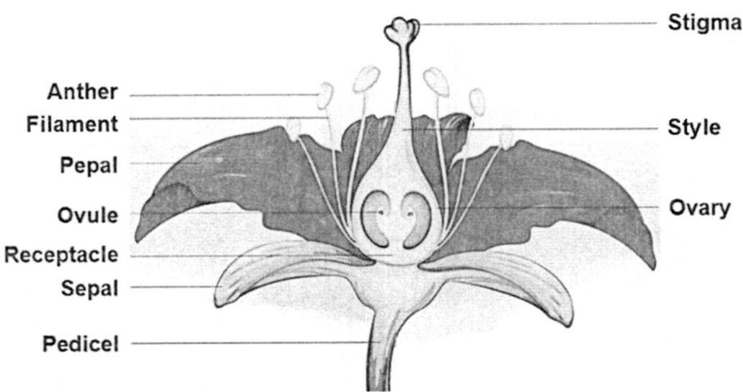

Fig. 16.5: Parts of flower

Whorls

Calyx is the outermost part of a flower. Sepals are the functional units of calyx, which protect the flower from desiccation or any other damage. ***Corolla*** is composed of petals. It is coloured to attract insects for pollination. It can be tubular, bell or funnel-shaped. ***Androecium*** is composed of stamens, the male reproductive part of a flower. It consists of a stalk/filament and anthers. ***Gynoecium*** is composed of one or more carpels, the female reproductive part of a flower. It consists of stigma, style and ovary. While stigma is the receptive surface for pollens, style connects the ovary to the stigma; ovary is an enlarged basal part of gynoecium. Each ovary has a placenta that consists of one or more ovules.

Aestivation

Aestivation is the arrangement of petals and sepals in floral buds before blooming. It is arranged with respect to the other members of the whorl (Figure 16.6).

Types of Aestivation

a. **Valvate**: It is an open aestivation having petals and sepals attached through the margins of a single whorl leaving a free space within them. Thus, petals and sepals of the whorl do not overlap each other. Such aestivation is found in the flower of Mustard and Calotropis.

b. **Twisted:** In case of twisted aestivation the petals and sepals are organized in such a manner that one edge overlaps the next petal/sepal edge inside. Thus, they overlap the neighbouring members on one side in a regular manner. Such aestivation is found in the flower of Ladyfinger, Cotton, and Hibiscus.

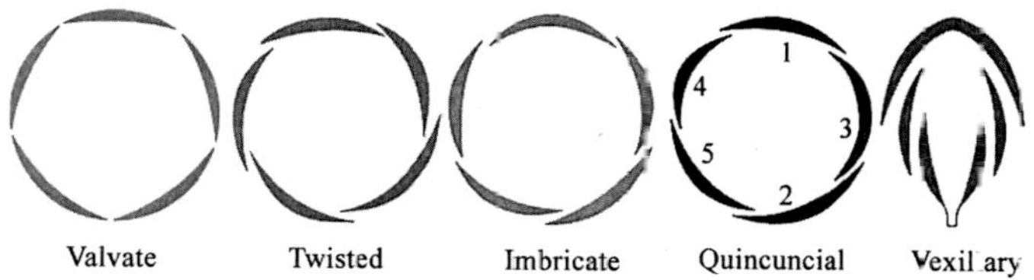

Fig. 16.6: Types of aestivation.

c. **Imbricate:** In this aestivation, the perianth or the outermost whorl, petals, and sepals overlap each other in such a way that a few petals are inside entirely and some are entirely outside. Such aestivation is found in the flower of Gulmohar.

d. **Quincuncial**: This is an example of ascending imbricate aestivation, wherein the petals start overlapping from the anterior end and continue till the posterior end. The perianth whorl is not arranged in a single whorl but is built in a spiral form. Thus, two petals are external, two are internal, and the fifth one has internal and external margins. Such an aestivation is found in the flower of Delonix and Cassia.

e. **Vexillary:** This type of aestivation (descending imbricate or papilionaceous aestivation) presents overlapping of petals from the bottom (posterior end) of the flower towards the anterior side. The overlap starts from the posterior side and ends at the anterior side. This is a special arrangement where one large petal covers the other smaller petals. A pair of the smallest anterior petals are overlapped by the lateral wings or the petals. Such aestivation is found in the flower of Pea, Crotalaria, and Dolichos.

Cohesion of Stamen

Cohesion of stamens can be of following five types (Figure 16.7):

a. **Monadelphous**: In this form of cohesion, all the stamens unite together in one bundle by the joining of filaments. Such cohesion of stamens is found in Hibiscus and Crotalaria.

b. **Diadelphous**: In this, stamens unite into two bundles. Such cohesion of stamens is found in Lathyrus and Pea.

c. **Polyadelphous**: In this case, there are numerous stamens and their filaments unite into a number of bundles. Such cohesion of stamens is found in Bombax and Citrus.

d. **Syngenesious**: Here, all the filaments are free but the anthers are united. Such cohesion of stamens is found in Sunflower.

e. **Synandrous**: When stamens are united by filaments as well as by anthers, it forms a compact body. Such cohesion of stamens is found in Cucurbits.

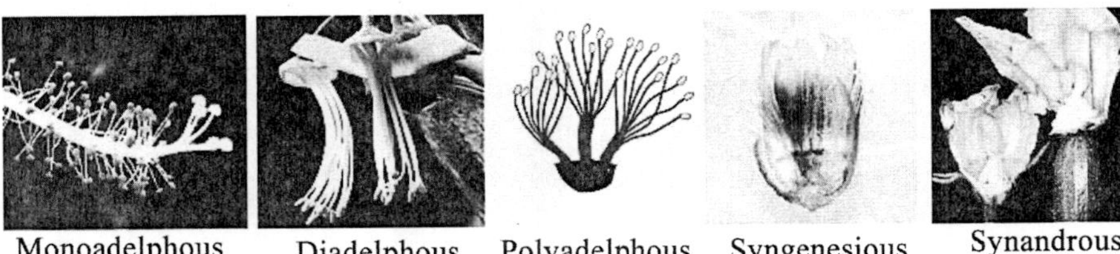

Fig. 16.7: Cohesion of stamens.

Inflorescence

Inflorescence is the arrangement of flowers on a floral axis. Different types of inflorescences are as described below:

1. Racemose

The main axis grows continuously, does not terminate into a flower, and produce flowers laterally in an acropetal manner (young flower on upper side while the older flowers on lower side). It can be of six types:

Raceme: Having elongated main axis and flowers are pedicellate. Such inflorescence is found in Mustard and Radish.

Spike: Having elongated peduncle but sessile flowers. Such inflorescence is found in Achyranthes.

Catkin: Having thin, weak, and long peduncle where flowers are sessile and unisexual. Such inflorescence is found in Mulberry and Betula.

Spadix: Having thick, long, and fleshy peduncle with small, sessile, and unisexual flowers. Such inflorescence is found in Colocasia and Maize

Corymb: Having short peduncle and flowers are present at the same level. Such inflorescence is found in Candytuft.

Capitulum: Here the peduncle doesn't grow and becomes broad, flattened concave or convex. Such inflorescence is found in Sunflower.

2. Cymose

In this case, the peduncle terminates in a flower; hence, the older flowers are at the upper portion of the plant while young buds are arranged towards the base in a basipetal manner. It can be of following three types:

Uniparous or Monochasial cyme: Here, peduncle terminates in a flower and generating lateral branch. It can be helicoid cyme and scorpioid cyme.

Biparous or Dichasial cyme: Here, peduncle terminates in a flower and the basal part produces two lateral branches.

Multiparous or Polychasial cyme: Here, peduncle terminates in a flower and the basal part produces many lateral branches, which also terminate with a flower.

Placentation

The pattern/arrangement of ovules within the ovary is known as placentation. There are following six major types of placentation.

1. **Marginal:** It has monocarpellary gynoecium and unilocular ovary. Such placentation is found in the members of Leguminosae or Fabaceae family, such as Pea, Gram, Mung, etc.
2. **Parietal:** In this case, polycarpellary gynoecium and carpels are united to form only one chamber. Under certain conditions, one chamber gets divided into two chambers due to the development of false septum. Such placentation is found in the members of Brassicaceae family, such as Mustard.
3. **Axile:** Here polycarpellary gynoecium is syncarpous and ovary possesses many chambers. Since the placenta originate from central axis, it is known as axile placentation. Such placentation is found in Petunia, China-rose, Tomato, and Lemon.
4. **Free Central:** Here polycarpellary gynoecium possesses unilocular ovary. There is no septum, only one swollen placenta bears a few ovules at the centre of the ovary, which is quite separate from the ovary wall. Such placentation is found in Dianthus and Primula.
5. **Basal:** In this case, polycarpellary gynoecium possesses unilocular ovary and a single ovule develops at the base of the ovary. Such placentation is found in the members of the family Asteraceae and Poaceae, such as Sunflower, Rice, Wheat, Maize, etc.

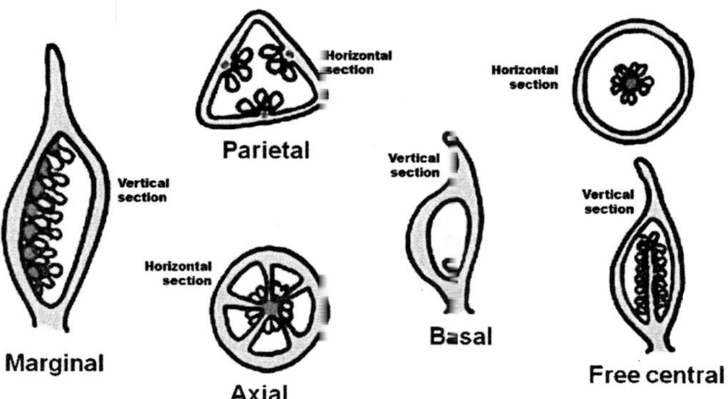

Fig. 16.8: Types of placentation.

6. **Superficial:** Herein, polycarpellary, syncarpous and multilocular ovary is found. Such placentation is found in Nymphaea (Waterlily).

C. Fruit

Fruit is the seed-bearing structure of flowering plants, which is formed from an ovary. After ripening, the ovarian wall changes into pericarp. This may be thick and fleshy, thick and hard or thin and soft. A pericarp has three layers namely:

- Epicarp: the outermost layer,
- Mesocarp: the middle layer,
- Endocarp: the innermost layer.

Types of fruits

a. **True Fruit**: it develops from the ovary only. Such examples are the fruits of Mango, Coconut, Zizyphus, etc.

b. **False Fruit or Pseudocarp**: It develops from other only parts than ovary like thalamus, inflorescence, calyx, which gets modified and become a fruit called false fruits. Examples of such fruits are Apple, Strawberry, etc.

Classification of Fruits

Fruits can be classified into four groups namely:

- Simple
- Accessory

- Aggregate
- Composite

1. ***Simple fruit:*** develops from a monocarpellary ovary or multicarpellary syncarpous ovary. Only one fruit is formed by the gynoecium. Simple fruits can be of two types:

   a. **Dry fruits**: the pericarp dries when the fruit matures. This can be (i) ***Indehiscent*** – pericarp does not split on maturity like **Achene**: one seeded, thin pericarp, seed coat is separate from ovary wall. An example is Sunflower seed. **Grain** or **Caryopsis**: which is one seeded, pericarp fused with ovary wall. An example is Wheat, Rice, Maize. **Nut**: is one seeded fruit with hard or bony pericarp. An example is Almond, Walnuts. (ii) ***Dehiscent*** – pericarp splits open to release the seeds on maturity, which might be a **Follicle**: having a pod like fruit that splits along ventral side e.g., lilac. **Legume**: a pod that splits along two opposite sides, e.g., Beans, Pea, Pigeon pea. **Silique**: wherein pericarp splits along the juncture of two carpels, e.g., Mustard. **Capsule**: having pericarp opening at pores along slits, e.g., Poppy, Bhindi. **Schizocarp**: the fruit slit in two or more achene like segment with the seeds surrounded by fruit wall, e.g., Coriander.

   b. **Fleshy fruits**: part or all the pericarp is fleshy on maturity. They can be **Drupe**: one/two seeds with fleshy mesocarp and stony endocarp, e.g., Mango, Plum, Peach, Coconut. **Berry**: one or several seeds, fleshy mesocarp, endocarp not stony, e.g., Grape, Tomato. **Hesperidium**: it is a special berry with a leathery rind and oil glands with dotted fruit surface, e.g., Citrus.

2. ***Accessory fruits:*** is berrylike, but produced from an inferior ovary; thus contain accessory structures. These may be **Berries**: for example Coffee, Banana. **Pepo**: with combined accessory tissues and ovary wall forming a tough rind, e.g., Cucumber, Watermelon.

3. ***Aggregate fruits:*** develop from multicarpellary, apocarpous ovary. Each carpel is separated from one another and makes fruitlet. These make a bunch of fruitlets known as etaerio. Etaerio can be of four major types:

   a. Etaerio of follicles: e.g., Calotropis, Magnolia

   b. Etaerio of achenes: e.g. Strawberry, Rose, Lotus.

   c. Etaerio of berries: e.g., Polyalthia, Custard-apple

   d. Etaerio of drupes: e.g., Raspberry.

4. Composite or Multiple Fruits

These are formed by fusion or collective shedding of fruits made from numerous independent flowers. These may be of two types. (i) **Sorosis**: develops from spike, spadix or catkin inflorescence, e.g., Jackfruit fruit. (ii) **Sycosis**: develops from hypanthodium inflorescence, e.g., Ficus fruit and Peepal.

Let's Check Your Understanding

1. Male reproductive part of flower is called ……
2. Mature ovary is termed as ………..
3. Major function of corolla is ………
4. What is aestivation in plants ……
5. Example of Etaerio of drupes is ……..
6. Type of placentation is found in mustard and other members of the Brassicaceae family is ………
7. Type of placentation found in rice and wheat is …………
8. Mango is a ……….. type of fruit.
9. Capitulum is a type of inflorescence where peduncle doesn't grow and becomes broad, flattened concave or convex.

   True / False
10. Sorosis type of fruits develops from which inflorescence?

Answers

1. Androecium
2. Fruit
3. To attract insects for pollination
4. Aestivation in Plants is the arrangement of petals and sepals in floral buds before blooming.
5. Raspberry
6. Parietal
7. Basal
8. Drupe (Fleshy)
9. True
10. Spike, Spadix or Catkin

17

Handling of Hazardous Materials

~Working with chemicals?

Reduce the risk to be safer!

Introduction

We come across a variety of chemicals every day in a laboratory. While some of them are relatively safer, others are extremely hazardous. Thus, it is very important to know the potential hazards that can be caused by the chemicals present in the laboratory and the safety measures needed to be followed to avoid the hazards. Additionally, we should also be aware of the steps necessary to be taken on such hazardous situations. According to IUPAC, a hazardous chemical mean the chemical which has the potential to cause damage in various ways. In a laboratory, there are chemicals that might be irritant, carcinogen, respiratory fibrogen, systemic poison, asphyxiant, flammable, etc. which have the potential to cause hazard to the workers or environment. Therefore, it is important to use appropriate procedure to reduce the risk and protect the health/safety of laboratory personnel, the public and the environment.

Types of Hazards Associated with Chemicals

Chemical hazards: The hazards caused by chemicals can be broadly classified into three categories.

1. Health hazard: This is also known as toxic effects of chemical. The risk associated with a chemical in laboratory depends on the extent of exposure and the inherent toxicity of the chemical. Many chemical reactions form the product that might be much more toxic than the reacting chemicals. For example, inadvertent mixing of formaldehyde (a common tissue fixative) and hydrogen chloride results in the generation of bis(chloromethyl) ether, a potent human carcinogen. For most of the chemicals, the route of exposure (through skin, eyes, gastrointestinal tract, or respiratory tract) is more important (Figure 17.1) that must be considered during risk assessment.

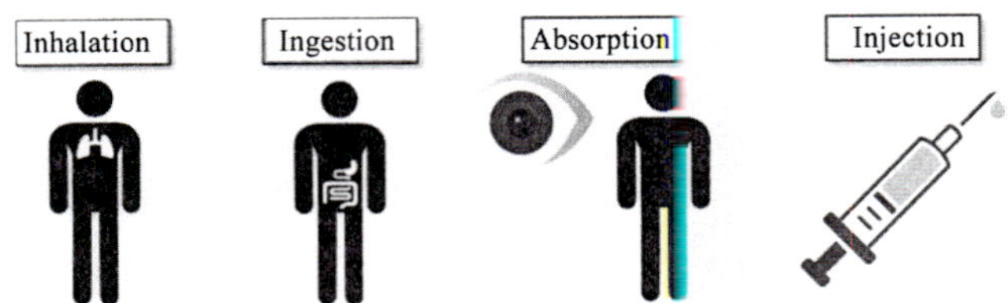

Fig. 17.1: Common routes of exposure to chemicals.

Laboratory chemical might be a toxicant, which may be an irritant e.g. formaldehyde, iodine; corrosive e.g. bromine, sulphuric acid, chlorine, nitrogen dioxide, sodium hydroxide, phosphorus; allergen e.g. diazomethane, dicyclohexyl carbodiimide, formaldehyde, beryllium, platinum etc.; asphyxiants e.g. acetylene, carbon dioxide, argon, helium, ethane, nitrogen, and methane. It might also be a neurotoxin e.g. mercury (inorganic and organic), organophosphatic pesticides, carbon disulfide, xylene, etc.; reproductive/developmental toxin e.g. ethylene glycol ethers (which may cause miscarriages), carbon disulphide, halogenated hydrocarbons, nitro aromatics, arylamines, ethylene glycol derivatives, etc.

2. Physical hazards: While working with chemicals in a laboratory, flammability, reactivity, and explosion hazards of chemicals should also be considered in addition to their toxic effects.

Flammable hazard: The substance that readily catch fire and burn in presence of air is known as flammable substance. These can be solid, liquid or gas.

Reactive hazard: There are chemicals that react violently with water and generate huge amount of heat and flammable gas upon reaction (e.g. alkali metals like lithium, sodium; organometallic compounds and hydrides). Some anhydrous metal halides (aluminium bromide), oxides (calcium oxide), and non-metal oxides (sulphur trioxide) react exothermically with water resulting in a violent reaction.

Pyrophorics: The materials that oxidize so rapidly in presence of oxygen or moisture that it cause ignition are known as pyrophoriocs. Examples are metal hydrides, low valent metal salts, iron sulphides, etc.

Incompatible chemicals: When two highly incompatible substances (e.g. strong oxidizing and strong reducing agents) are mixed accidentally, it may lead to explosion or the formation of toxic/ flammable substances.

3. Ecological hazards: Such hazards are caused by contamination of environment with the hazardous chemicals being used by the workers for research, industrial, medical purposes, etc.

Principles of Handling Hazardous Chemicals

There are four fundamental principles which should be considered before starting work in a laboratory.

i) Planning ahead: Determine the potential hazards associated with the use of chemical before starting the experiment.

ii) Minimize exposure: Do not allow laboratory chemicals to come in contact of skin. Use hoods and other ventilation devices to prevent the exposure.

iii) Do not underestimate hazards: Treat all compounds/substances of unknown toxicity as toxic.

iv) Be prepared for accidents: Be prepared for the specific action to be taken in the event of accidental release of a hazardous substance.

Table: Classification of hazardous materials, their symbols, and examples.

Class	**Hazardous material**	**Examples**	**Symbol**
I	Explosives	Ammunition, Fireworks, Air bag inflators	EXPLOSIVES 1
II	Gases	Fire extinguishers, Propane, Carbon dioxide	FLAMMABLE GAS 2; NON-FLAMMABLE GAS 2; INHALATION HAZARD 2; OXYGEN 2
III	Flammable liquids	Paints, Alcohols, Gasoline	FLAMMABLE 3; GASOLINE 3; FUEL OIL 3; COMBUSTIBLE 3
IV	Flammable solids, spontaneously combustible	Matches, Sulfur, Oily fabrics	

Class	Hazardous material	Examples	Symbol
V	Oxidizer and hydrogen peroxide	Hydrogen peroxide, Chlorates, Ammonium nitrate fertilizers	OXIDIZER 5.1; ORGANIC PEROXIDE 5.2; ORGANIC PEROXIDE 5.2
VI	Toxic, inhalation hazard and infectious substances	Medical wastes, Dyes, Biological specimens	POISON 6; TOXIC 6; INHALATION HAZARD 6; INFECTIOUS SUBSTANCE 6
VII	Radioactive	Uranium	RADIOACTIVE 7
VIII	Corrosive	Acids (Sulfuric acid), Iodine, Batteries	CORROSIVE 8
IX	Miscellaneous	Dry ice, First aid kits	9

SOP for Handling of Hazardous Chemicals

The following Standard Operating Procedure set for a laboratory must be remembered while working with hazardous chemicals.

a) **Assess the known and potential hazards:** Prior to start an experiment, the laboratory staff must evaluate the potential hazards of using the chemicals. Material safety data sheets and label on the container as well as other references might be used for the assessment. The process should include preparation for any unexpected/potential emergency.

b) **Substitution of more hazardous by lesser one:** If possible, it is safer to consider substituting a more hazardous chemical with less hazardous/ toxic substance. Moreover, only those chemicals for which appropriate exposure/emergency control measures are known/present in the laboratory must be used.

c) **Engineering and administrative controls:** Laboratory staff not involved in handling of chemicals must be kept away from the chemicals. Those who wish to work with must obtain prior approval to proceed with a laboratory task from the supervisor/engineer concerned whenever he/she wish to perform a new laboratory procedure/test, a significant change in a procedure/test likely to alter the hazard or having unknown/ unexpected test results.

d) **Reporting incidents/unsafe conditions:** Without considering the level of incidence (minor or major), any casualty happening in the laboratory must be reported to the Lab Supervisor. The following unsafe conditions in a laboratory must be reported.

- Non-functioning of hoods dedicated for the handling of hazardous chemicals must be informed,
- Unsafe storage conditions in the laboratory,
- Any blockage of emergency exits from laboratory,
- Improperly charged fire extinguishers;
- Not functioning of eyewash stations or safety showers;
- Limited stock/unavailability of personal protective equipment.

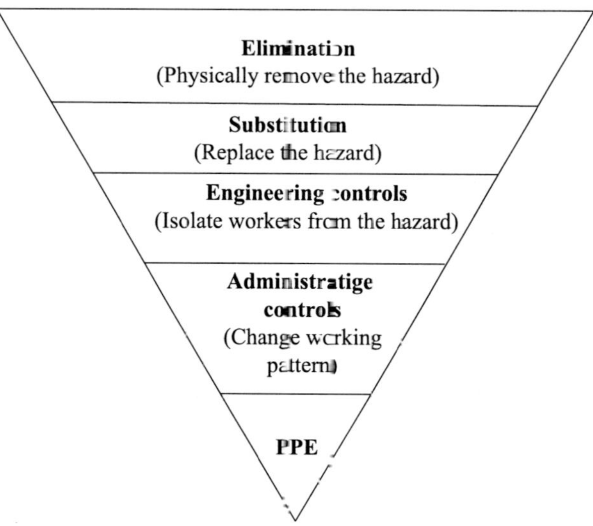

Fig. 17.2: SOP for handling hazardous chemicals.

General Guidelines for Safe Handling of Chemicals

1. Always label container with name of the chemical.
2. Use protective equipment for eyes and make sure to wear lab coat while handling chemicals.
3. Avoid smelling, inhaling, and tasting chemicals.
4. Avoid touching chemicals using hands, face, clothes or shoes.
5. Hazardous chemical should be used following appropriate instructions as directed.
6. Read the warning label before opening reagent bottle. This will help knowing the appropriate/ special (refrigeration or inert atmosphere) storage precautions.
7. A periodic check on chemical containers (for rust, corrosion, and leakage) must be performed.
8. Smoking, drinking, eating, and application of cosmetics must be forbidden in the areas where hazardous chemicals are used or stored.
9. Before leaving the lab after handling chemicals, hands must be washed thoroughly with soap and water.

Personal Hygiene in Laboratory

- Never store food or beverages in chemical storage areas, refrigerators, glassware or elsewhere in a laboratory.
- Never taste or smell chemicals.
- Before leaving laboratory, wash hands/areas exposed to chemicals.
- Confine long hair and loose clothing, wear shoes all the times being in laboratory; do not wear sandals, perforated shoes or sneakers.
- No wearing of jewellery (e.g. rings) is allowed in a laboratory, as it may interfere with gloves and other protective clothing, may come in contact with electrical supplies or reacts with chemicals.

Safety Data Sheets

Safety Data Sheet (formerly MSDS) provides proper procedures for handling the chemical as well as emergency measure to the users of hazardous material. It includes the information on:

- Health, physical and ecological hazards associated with the material/ chemical,
- Physical properties, reactivity, and toxicological data for the chemical,
- Measures like first aid, storage, disposal, exposure control, and spill/leak procedures, etc.

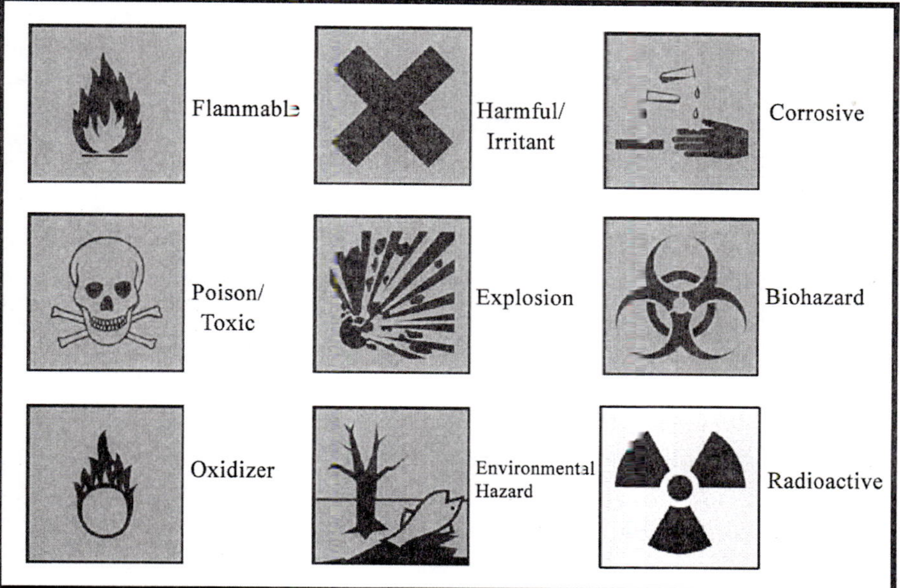

Fig. 17.3: Must know the meaning of signs.

Chemical-specific Safety Measures

a) **Handling of flammable chemicals:** Flammability of chemical is determined by its flash point, the temperature at which an ignition source can cause the chemical to ignite momentarily under certain conditions. Chemical with a flash point below 93.3°C is considered fire-hazard chemical. Fire-hazard chemical must be stored in a flammable storage room or in flammable storage cabinets, must be used only in chemical hoods, kept away from sources of ignition. When working with flammable chemicals, be certain that there are no sources of ignition nearby, which may cause fire or explosion.

b) **Handling of pyrophoric chemicals:** All pyrophoric chemicals should be stored in tightly closed containers and when being transferred, they should be kept under an inert atmosphere/liquid.

c) **Handling of reactive/unstable chemicals:** To minimize decomposition, peroxides must be stored at lowest possible temperature consistent with their solubility/freezing point because peroxides at higher temperature are extremely sensitive to shock and heat.

d) **Handling of water-reactive chemicals:** Water-reactive chemicals must be kept away from water or water-containing materials.

e) **Handling of corrosive/contact hazard chemicals:** The corrosive and contact-hazard chemicals must be handled with proper safety and precautions including the use of safety goggles and face shield, gloves resistant to permeation/penetration, and laboratory coat. Corrosive chemicals that may generate corrosive gases, vapours, fumes, aerosols, and dusts must be used in hood or other containment device.

f) **Handling of solid chemicals:** A chemical even in solid form might be equally hazardous. Individuals sensitive to nickel may develop dermatitis on contact with skin. A fuming solid may emit toxic vapours that might be inhaled by the individuals in sounding areas. While transferring solid chemicals use of spatula is recommended; however, you should make it sure that the spatula is clean and dry. Never use the same spatula for different solid chemicals, unless it is cleaned and dried. If solid chemicals need to be mixed, place the chemicals on separate pieces of paper or in plastic weighing boats and then pour one solid chemical gently on another. By repeat the pouring action (back and forth) about 10 times the chemicals can be thoroughly mixed. This is the safer way of mixing explosive chemicals, which might otherwise explode if mixed using a metallic spatula.

Type of Spatula

Laboratory spatulas are the tools that help in mixing, scraping, and other tasks related to transfer of solid chemicals/samples from one place to another. They are used for transfer of chemicals from their original containers to weighing paper, weighing boats, weighing bottles, weighing funnels, or other vessels while weighing. Reusable spatulas are generally made of stainless steel, as they are resistant to heat, cold, acids, bases, solvents, and other chemicals. Metal spatula may also have removable or permanent polymer (PTFE or plastic coated) or wooden handle. Spatulas of different types available for various work include:

Microspatula having pointed ends that help transferring materials to smaller tubes or microplates. ***Trough-shaped spatula*** has one rounded and one pointed end, which is used to dislodge/loosen chemical in a container. ***Double-ended spatula*** has both the ends of equal or different sizes/shapes. This may have *one flat and one bent end, one rounded and one tapered end, one rounded and one scoop/spoon-shaped end.*

Disposable plastic or paper made spatulas are also used in certain laboratories. Disposable spatulas help eliminate the chances of cross-contamination, which might occur when poorly cleaned reusable spatula is used repeatedly. For performing lyophilization or handling lyophilized product, generally a disposable plastic spatula is recommended, as this is less affected by low temperature.

Fig. 17.4: Different types of spatulas.

Antidote to Counteract Poisoning

Antidote is an agent that can negate the effect of a poison/toxin. An antidote mediates its effect either by preventing the absorption of the toxic substance, by binding and neutralizing the poison, antagonizing its effect, or inhibiting conversion of the toxin to more toxic metabolite(s). It is prudent to presume that every chemical in a laboratory is toxic/poisonous hence they must be handled with appropriate care with safety. In case of any mishappening/abuse of chemicals. The following first-hand treatment must be practiced until the affected person reaches appropriate health centre for necessary treatment.

- In case a chemical is swallowed, immediate action must be taken to treat the affected person.
- In case acetone is consumed, vomiting should be induced and a universal antidote should be administered while keeping the patient awake.
- Universal antidote can be prepared by mixing 2 parts of pulverised charcoal, 1 part of magnesium oxide and 1 part of tannic acid. A few grams of this mixture can be added to a cup of warm water and then administered. This universal antidote is administered when any other specific antidote is not known/available.
- In case acid is consumed, vomiting should not be induced but magnesium oxide or lime water must be administered immediately. Repeat this every 15–20 minutes. A few mL of mineral oil can also be administered four to five times every 15 minutes. Give milk or egg-white along with cold water. However, carbonates should not be given to the affected person.
- In case an alkali is consumed, administer 5% acetic acid or vinegar but do not induce vomiting. Administer milk or egg-white.
- In case methyl alcohol is consumed, induce vomiting and then administer milk or egg-white. Small amount of ethanol can be given to minimize the harmful effects of methanol on eyes.

First-aid Techniques

- In case an organic substance falls on skin, wash it off liberally with rectified spirit, then with soap and warm water.
- In case an acid falls on clothes, apply dilute ammonium hydroxide to the affected area and then wash thoroughly with water.
- In case an alkali falls on clothes, apply dilute acetic acid on the affected portion and then wash with plenty of water.

Let's Check Your Understanding

1. What is the flash point for flammable liquids?

   a. Above 93 °C b. Below 93 °C

   c. Above 100 °C d. None.

2. Chemicals which cause visible destruction or irreversible alterations to living tissues at the site of contact are known as?

   a. Inflammable chemicals b. Corrosive chemicals

   c. Oxidizing chemicals d. Sensitizers

3. Which of the following is not an example of flammable chemical?

   a. Gasoline b. Acetone

   c. Ethanol d. Nitrogen

4. Following symbol denotes the warning of

   a. Corrosive b. Radioactive

   c. Oxidizer d. Explosion

5. Chlorates and ammonium nitrate fertilizers comes under which class of hazardous materials?

   a. Gases b. Corrosives

   c. Oxidizers d. Radioactive

6. What is the most common route of exposure?

   a. Inhalation b. Ingestion

   c. Contact with skin and eyes d. Injection

7. Which of the following is not a chemical-related health hazard?

   a. Carcinogenicity b. Reactivity

   c. Corrosivity d. Toxicity

8. An acute health effect is

   a. A state of deep stupor produced by a chemical substance

   b. Adverse effects resulting from long term exposure to a substance

   c. Adverse effects resulting from a single short-term exposure to a substance

   d. Exposing a chemical substance to the environment

9. Abbreviation SDS stands for ………………

10. Abbreviation PPE stands for ………………

Answers

1. a
2. b
3. d
4. a
5. c
6. a
7. b
8. c
9. Safety Data Sheet.
10. Personal protective equipment.

18

Electrical Wiring and Earthing in Laboratory

~Knowledge of basics might minimise the chance of accidents

Introduction

Proper electrical wiring and earthing are crucial for laboratory safety and functionality. A laboratory houses several installations/equipment that demand stable/continuous power supply. Inappropriate or faulty wiring and earthing may lead to significant risks, including electrical shocks, damage to equipment, and even incidence of fire. Therefore, understanding the requirements of electrical wiring and earthing is essential for designing, construction, and maintenance of laboratory. This chapter aims to provide a basic knowledge of electrical wiring and earthing necessary for a laboratory. Importance of earthing in mitigating electrical hazards and ensuring integrity of electrical installations are also described.

Basics of Electrical Wiring

Basic understanding of electrical wiring is necessary for laboratory safety and efficient electrical functioning. The potential difference between two electrical points (negative and positive) is known as ***Voltage***, which drives the flow of electric current. This is measured as volt (V). The rate of flow of electricity in a circuit is known as ***Current*** *(I)*, which is measured as ampere (A). Current can be either ***direct*** (DC) or ***alternating*** (AC). Opposition/resistance experienced in the flow of current in circuit is known as ***Resistance*** *(R)* and it is measured in ohms (Ω). Higher the resistance means lesser the current flows at a given voltage. The rate at which electrical energy is transferred/utilized in an electric circuit is known as ***Power*** *(P)*, which is measured in watt (W). The Power can be calculated as the product of voltage and current ($P = V \times I$).

Type of Electrical Circuit

Series circuit contains different components connected end-to-end manner so that same current flows through each component; however, the voltage is divided among the components. It is simple in design/wiring, but when one component fails, the entire circuit is interrupted.

Parallel circuit contains different components connected across the same two points, each receiving the same voltage but carrying different currents. Each component operates independently, and failure of any one does not affect others. However, it is more complex in wiring compared to a series circuit.

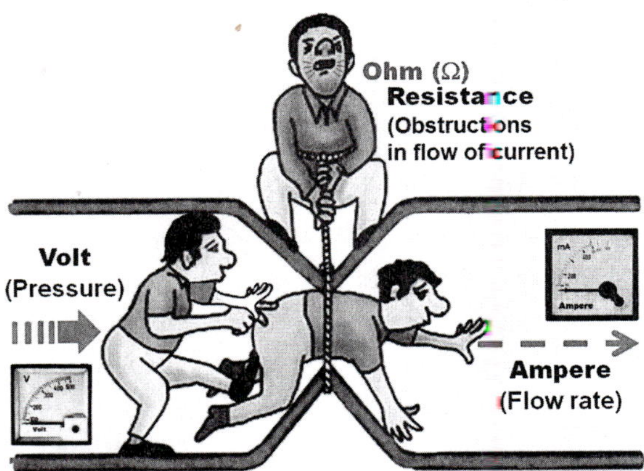

Fig. 18.1: Pictorial representation of the relationship between Current (Amp), Resistance (Ω), and Voltage (V).

Complex circuit contains a combination of series as well as parallel circuits, which is used to optimize the performance and ensure reliability in complex applications.

Types of Wiring

Conduit wiring involves electrical wires running through metal or plastic pipes (conduits) to protect them from physical damage and environmental factors. While *rigid conduit* offers excellent protection (but more difficult to install), *flexible conduit* is easier to install but provides less protection than rigid conduit. *Trunking* uses protective enclosures (trunks) to house multiple wires, allowing easy access, maintenance and modification whenever needed. It facilitates easier change/addition to the wirings. However, it may be bulky and visually intrusive. *Surface wiring* involves the wires running on the surface of walls/ceilings, often encased in protective channels or raceways. It is easier to install and modify but it is aesthetically less pleasing and potentially more susceptible to damage.

Electrical Codes and Standards

International Electrotechnical Commission (IEC) provides international standards for electrical, electronic, and related technologies, ensuring global harmonization and safety. *Institute of Electrical and Electronics Engineers (IEEE)* provides standards and guidelines for various electrical and electronic systems to ensure reliability and safety. *Laboratory-specific standards* may need to comply with additional standards related to the specific nature of the work.

Components of Electrical Wiring

Understanding different components of electrical wiring is necessary for making a laboratory safe and efficient. Copper, being highly conductive and durable, is widely used in electrical wiring. Although aluminium is less conductive than copper, being lighter and less expensive, it is often used in large-scale electrical applications. The cables used in wiring might be *Flexible cable*, used for the places where wires need to bend frequently, connecting equipment/devices or *Armored cable* having an additional layer of protection (usually metallic) to prevent physical damage, which is suitable for harsh environments or underground installations.

Switches, Sockets, and Outlets

Toggle switch is a commonly used switch to open or close the circuit in smaller equipment. *Rocker switch* operates by rocking a switch from one position to another position, and commonly used switch on medium sized equipment. *Push-Button switch* activates or deactivate circuit with a push button and it has been used earlier in smaller equipment. *House-hold switch* is generally used to switch ON or OFF the lights and fans in houses/laboratories. *Door-well switch* is used at outside the door for call bell (Figure 18.2).

Outlets and Sockets

Standard outlets provide connection points for electrical devices, which are available in different configurations to match different plug types and voltages. Ground Fault Circuit Interrupter (GFCI) outlets are designed to protect against electric shock by quickly cutting off power when a ground fault occurs. *USB outlet* provides standard power as well as USB charging ports for modern devices.

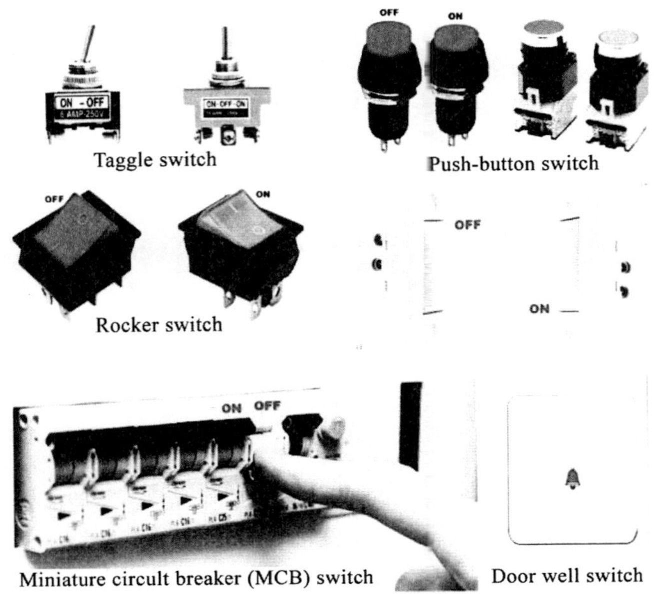

Fig. 18.2: Different types of switches used in electrical wiring

Circuit Breaker and Fuses

Miniature circuit breaker (MCB) automatically trips to cut-off the power supply when a fault, overcurrent or short-circuit occurs (Figure 18.2). *Residual current circuit breaker (RCCB)* detects difference between live and neutral currents to protect against electric shock and leakage of current.

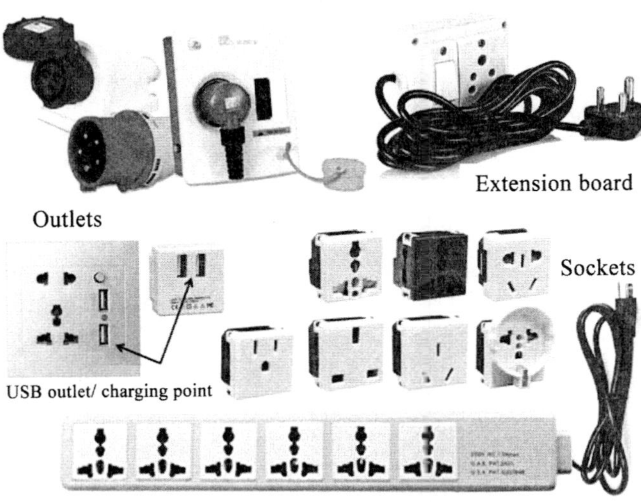

Fig. 18.3: Different types of outlets and sockets

Wire fuse contain a wire that melts when too much current flows through it, thereby interrupts the circuit, which can be rewired/replaced after correcting the problem. *Cartridge fuse* contains a wire or metal strip that melts when too much current flows through, thereby interrupting the circuit (Figure 18.4).

Electrical Panel and Distribution Board

Electrical panel is the central hub for distribution of electricity to different building/houses or facility. It houses main switches, circuit breakers, and safety devices. *Distribution boards* are the sub-panels that distribute electricity to specific areas or systems within a building. It facilitates management and segregation of different phases for maintenance and safety reasons.

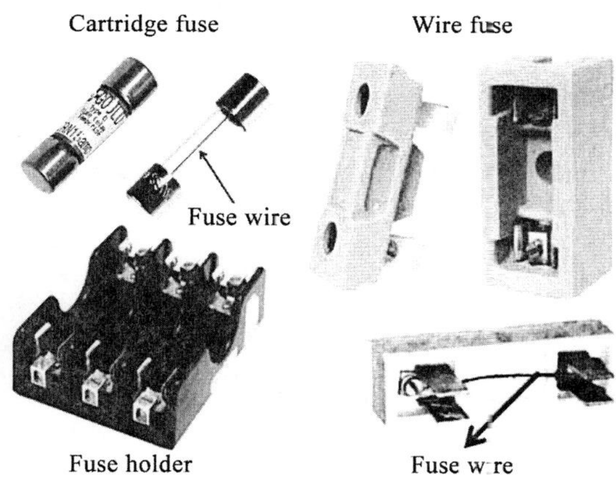

Fig. 18.4: Different types of fuse

Grounding and Bonding Components

Grounding rods are metallic bars driven into ground to provide a direct path to the earth for faulty currents. Grounding prevents damage to equipment and the installation. *Grounding conductor* is the wire that connects body of the electrical equipment to the ground/earth to neutralizing/safe dissipation of any faulty electrical current leaking to body of the equipment; thereby ensuring safety to the user against electrical shocks. *Bonding jumpers* ensure electrical continuity between metallic parts and conductive materials, reducing the risk of potential differences that can lead to shocks or fires. Copper being a common choice for grounding rods, other materials such as aluminium, zinc, stainless steel, copper-coated galvanized steel, etc. can also be used as grounding rod and earthing plate. However, it depends on the local code/ regulations, cost considerations, and specific application/requirement. Due to excellent electrical properties, low resistance, good conductivity, resistance to corrosion, long life expectancy, etc. solid copper is a popular/reliable choice for grounding and earthing systems.

Electrical Design and Layout of Laboratory

Careful planning, designing and layout of electrical systems in laboratories are essential for efficiency, safety, compliance with relevant standards, and minimize the risk of accidents.

Electrical Systems for Laboratories

Chemistry laboratory: Requires robust electrical systems to power fume hoods, spectrophotometers, and other sensitive equipment. It must include explosion-proof outlets and switches, particularly in areas where volatile chemicals are used.

Biology laboratory: Needs stabilized power for incubators, centrifuges, and microscopes. Undisrupted power backup might be crucial for the experiments and the equipment present in the laboratory.

Physics laboratory: Demands precise voltage regulation for sensitive experiments and equipment. Often require customized wiring/layouts to accommodate specialized apparatus.

Load Assessment and Electrical Distribution

Calculation of load: Assess the required electrical load by summing power ratings of all equipment and devices. Also consider future expansions and additional equipment required in the laboratory to avoid overloading circuits in the laboratory.

Distribution of load: Distribute the load evenly across multiple circuits to prevent overloading. Use dedicated circuits for heavy/high-power equipment to ensure stability and safety.

Placement of electrical outlets and switches: Position outlets near workstations and equipment to minimize the use of extension board. Ensure easy access to switches for controlling equipment and lighting.

Safety Considerations

Install GFCI outlets in wet or damp areas to protect against electrical shocks. Use tamper-resistant outlets in areas accessible to student and trainees.

Emergency Power Supply

Uninterruptible power supply (UPS): Provides power supply without any interruption during electricity outrage to avoid any disturbance in the on-going work, protect sensitive equipment, as well as the data. Continuous electrical power supply/backup can also be ensured by using a *Generator*, which offers long-term power backup against extended outage. However, it needs to be of appropriate size (as per the requirement of the laboratory/building). An automatic transfer switch can be used to switch the power supply to the backup source on electrical outage.

Installation of Electrical Wiring

Installing electrical wiring in laboratory setup requires precision and adherence to safety protocols. It requires step-by-step process to install wiring along with the necessary tools, equipment, and safety precautions. *Planning and design* begin with laying out detailed plan and scheme of wiring. It needs to ensure that all the components and materials meet the required standards and specifications. *Rough-in wiring* begins with installing conduit, boxes, and supports as per the design. Wires are placed inside conduits or raceways, ensuring their proper labelling and organization. *Final wiring* wires are connected to switches, outlets, and fixtures. Colour-code must be followed following the conventions to maintain consistency and safety. *Testing and verification* is needed for each circuit to ensure continuity, insulation, and grounding.

Tools and Equipment

Basic tools required during wiring include screwdrivers, pliers, wire strippers, and cutters. Measuring tape, levels, and marking tools might also be required for specific purposes. *Specialized tools* like conduit benders, fish tapes, and pulling equipment are required for running wires through conduits. Multimeters and insulation testers are required to verifying electrical parameters. *Safety gear* like insulated gloves, safety glasses, protective clothing (Figure 18.5), ladder/scaffolding should be used for safety purposes and accessing areas located at height.

Safety Precautions

Personal safety measures require turning off power supply at the main panel before starting any electrical work. Use of insulated tools and wearing protective gear to prevent electrical shocks are necessary part of personal safety. *Worksite safety* requires keeping the work area clean and organized to prevent any accident. *Electrical safety* requires following lockout/tagout procedures to ensure circuits are de-energized during electrical work/repairing. Working in wet or damp conditions must be avoided to prevent electrical hazards.

Fig. 18.5: Components of safety gear

Testing and Verification of Wiring

Continuity testing: it ensures that there is no break or fault in wiring. A continuity tester or multimeter is used to verify the connection. *Insulation resistance testing:* checks for adequate insulation which is needed to prevent short-circuit and leaks. *Grounding verification:* confirms that grounding connection is secure and effective. It measures the resistance of grounding system to ensure that it meets the standards. *Functional testing:* is performed to test outlets, switches, and fixtures to ensure that they operate correctly by simulating different load conditions to verify their performance.

Common Wiring Issues

Short circuit occurs when a live wire comes in direct contact of another wire leading to excessive current flow. *Overload* results due to drawing more current than the circuit is designed to handle, which may lead to overheating and short-circuit. *Loose connection* may cause power fluctuation, intermittent faults, sparking, and heating.

Earthing System

Earthing or grounding is essential in laboratory to ensure safety and prevent any electrical hazard. *Safety* is necessary for personnel protection from electric shock by providing a path for faulty current to pass to the ground. It is also necessary for *protection of equipment* by dissipating faulty current. *System stability* requires proper earthing, which is necessary for maintaining consistent voltage supply. There are different types of earthing system which include *Terre Neutral (TN) system*, wherein neutral point of power supply is directly earthed. Its subtypes are TN-S (separate protective earth and neutral conductors), TN-C (combined protective earth and neutral conductors), and TN-C-S (combined and separate conductors). *Terre Terre (TT) system* has its own separate earthing connection, independent of the earthing of the power supply system. Moreover, in case of *Isolated Terre (IT) system*, the supply system is isolated from earthing or it is connected through a high impedance (with the installation earthed separately).

Components of Earthing System

Earthing rod and plate are metal parts buried in ground to provide a low-resistance path for faulty currents. *Ground conductor* is a wire connecting the electrical system to earthing rod/plate. *Earth busbar* is a central point in the electrical system where all ground conductors are connected. *Bonding jumpers* ensure electrical continuity between conductive parts and equipment. The methods of earthing include *Direct earthing* wherein the equipment is directly connected to the earth using earthing rod/plate. In case of *Indirect earthing* the equipment is connected to the earth through a resistor/impedance to limit faulty current.

Testing and Maintenance of Earthing

Earth resistance testing is performed to measure the resistance between the earthing electrode and the ground to ensure it meets the safety standards. *Continuity testing* ensures that all connections in the earthing system are intact and secure. *Regular inspection* is required for periodic checking of the integrity and performance of the earthing system.

Protection Systems

Protection systems are essential to safeguard electrical systems in laboratory from faults and ensuring safety of the laboratory personnel. Surge protection devices include *Surge arrester* that protects electrical equipment from voltage spikes by diverting surge currents to ground. *Transient voltage suppression* protects sensitive electronic circuits from transient voltage spikes. Whereas, *Ground fault circuit interrupter* detects ground faults (leaking currents) and interrupts the circuit to prevent electric shock. It is used in wet or damp areas like the places near sink and laboratories where liquids are used.

Lightning protection system uses *Lightning rod* to provide a direct path for lightning currents to the ground, protecting buildings and equipment from lightning strikes. *Surge protector* is installed in power lines to provide protect against lightning-induced voltage surge.

Regulatory Requirements for Electrical System

It must be ensure that all installations in a laboratory comply with national and local electrical codes and standards. Necessary certifications from regulatory bodies must be obtained to ensure safety. Detailed records for electrical installations, inspections, and maintenance activities must be maintained for regulatory purposes.

Common Issues and Troubleshooting

Identifying and resolving common electrical issues is crucial for maintaining safe and efficient laboratory environment. *Earthing issues* include *high earth resistance* due to poor grounding, which can be resolved by improving the earthing connection to ground. *Faulty earth connection* can be repaired by replacing damaged or corroded components.

Let's Check Your Understanding

1. What is the primary purpose of earthing in laboratory electrical system?
   a. To increase the voltage
   b. To protect personnel from electric shocks
   c. To reduce power consumption
   d. To enhance lighting efficiency
2. Which type of cable is commonly used in the environment where wires need to bend or move frequently?
   a. Armored cables
   b. Flexible cables
   c. Coaxial cables
   d. Ribbon cables
3. In a TN-C-S earthing system, what does the 'C' stand for?
   a. Combined
   b. Circuit
   c. Conduit
   d. Copper
4. Which device provides protection by interrupting the circuit when it detects a ground fault?
   a. Surge protector
   b. Circuit breaker
   c. Ground Fault Circuit Interrupter (GFCI)
   d. Lightning rod
5. Which component in an electrical wiring system serves as a central hub for distributing power to various circuits?
   a. Conductor
   b. Circuit breaker
   c. Electrical panel
   d. GFCI outlet
6. Which international organization develops standards for electrical and electronic technologies?
   a. NEC
   b. IEC
   c. OSHA
   d. UL

7. What should be done before starting work on an electrical installation to ensure personal safety?
   a. Turn off power at the main panel
   b. Increase the circuit's voltage
   c. Connect all wires to the same terminal
   d. Remove all fuses
8. Which of the following is NOT a type of electrical circuit?
   a. Series circuit
   b. Parallel circuit
   c. Complex circuit
   d. Direct circuit
9. What is used to protect against voltage spikes by diverting surge currents to the ground?
   a. Circuit breaker
   b. Surge arrester
   c. GFCI outlet
   d. Fuse
10. What is the purpose of a routine insulation resistance test?
    a. To increase electrical load
    b. To ensure adequate insulation and prevent short circuits
    c. To verify the color-coding of wires
    d. To calibrate electrical devices

Answers

1. b
2. b
3. a
4. c
5. c
6. b
7. a
8. d
9. b
10. b

19

Safety Measures in a Laboratory

~Safety must get priority over accomplishing research!

For a researcher, laboratory is another home where he/she spends much time every day. As we take care of cleanliness, safety, and design of our home, we must take care on these aspects of laboratory too. It is the responsibility of the laboratory personnel to ensure cleanliness/safety of the laboratory. Since every laboratory chemical and equipment have the potential to cause harms if not used with appropriate precautions. Hence, as a laboratory user, it is our responsibility to follow the basic safety/biosafety rules and regulations set up by the Principal Investigator of the laboratory, Institutional Biosafety Committee, as well as by other regulatory authorities, if any. We must be aware of the general/personal protective measures, cleanliness and laboratory waste-disposal prior to joining a research laboratory. In this chapter we will briefly discuss about personal safety, chemical safety, cleanliness, waste disposal, etc.

A. Personal Safety

1. *Personal Protective Equipment (PPE)*

Personal protective equipment (PPE) is special arrangement to protect the worker from specific hazards because of using a hazardous substance. PPE does not reduce or eliminate the hazard, but protects the wearer from exposure to hazardous material. PPE should not be reused or brought outside the laboratory as these might increase the risk of spread hazardous material or contamination. Used PPE must be kept inside the laboratory in a designated waste storage area adjacent to the work area.

Protective Clothing

Protective clothing provides a physical barrier between the chemical/hazardous material and the worker, which might be disposable or reusable. The type of protective clothing must be based on the level of risk involved. The reusable clothing typically includes laboratory coats, goggles, face shield, etc. Laboratory coat offers convenient and low-cost protection (Figure 19.1). This can be easily stored when not in use and can be washed/cleaned frequently. Apron, usually made of cloth (usually tied around the waist to protect clothing from harmful chemical/substance) is reusable, while those made of plastic or leather is typically impermeable and can be difficult to decontaminate. Such disposable laboratory protective wear/clothing typically consists of

- Hair and shoe covers,
- Synthetic over-garments,
- Protective sleeves, etc.

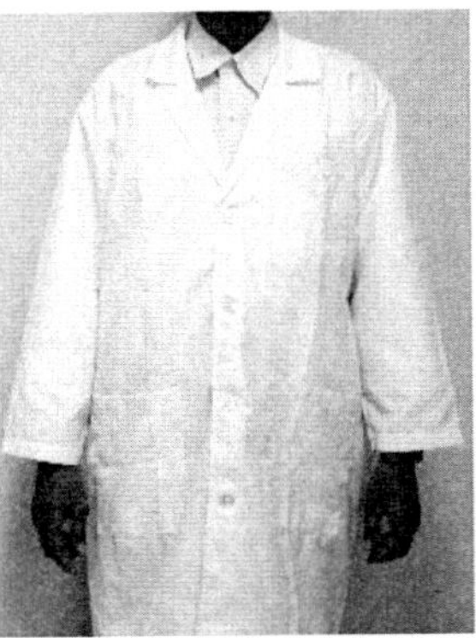

Fig. 19.1: Lab coat/Apron.

Disposable clothing eliminates the need for cleaning/washing and post-use storage. Additionally, it significantly reduces the possibility of secondary contamination, provides impermeable protection against liquids and offers better protection from biohazardous agents. Disposable clothing, however, may be costly for many laboratories. Loose clothing (such as overlarge clothing or tie), skimpy clothing (such as shorts), torn clothing and unrestrained hair may increase accidental hazards in a laboratory.

Gloves

Gloves provide primary protection to the hands of the laboratory worker to exposure from hazardous chemicals. Gloves must be worn in hands whenever there is a risk of contact with hazardous materials and the gloves should never be reused. Latex and Nitrile are the most common used gloves in laboratories. A variety of gloves do exist for different applications, however, while selecting a particular type of glove, the following factors must be considered:

- Allergies of laboratory personnel
- Length of cuff
- Thickness of material
- Use of chemicals

Gloves must be changed whenever a spill occurs or when a breach in the glove material is discovered. The use of gloves can be eliminated when transporting materials outside of the laboratory if the material is placed in a secondary container and the outer surface is decontaminated.

Many chemicals can not only cause skin irritation or burns, but also get absorbed through the skin, which can be a significant route of exposure to harmful chemicals. Dimethyl sulfoxide (DMSO), nitrobenzene, and many solvents are examples of such chemicals that can be readily absorbed through skin into the bloodstream and may cause harmful effects.

How to Select Gloves Based on the Material?

When selecting the glove, the following characteristics of its manufacturing material should be considered: *Degradation* of gloves/material is one of the physical properties need to be considered, as glove may be degraded on contact with particular chemicals. Degradation of gloves may appear as hardening, stiffening, swelling, shrinking or cracking on the gloves on contact with a chemical. Extent of degradation indicates how well gloves can protect worker's hand on exposure to a chemical. *Breakthrough time* is the elapsed time between the initial contact with a chemical on the surface of glove and the analytical detection of the chemical on the inside of the glove. *Permeation rate* is the rate at which a chemical passes through the glove/material on breakthrough/contact and equilibrium reaches. Permeation and degradation tests are performed under laboratory conditions, which vary significantly from actual laboratory conditions. It is recommended that the glove material should be selected based on the shortest breakthrough time. The following table provides information on the major glove types and their general uses; however, the list is not exhaustive.

Glove material	General uses
Butyl	Offers the highest resistance to permeation by most gases and water vapour. It is especially suitable for the usage of esters and ketones.
Neoprene	Provides moderate abrasion resistance but good tensile strength and heat resistance. It is compatible with many acids, caustics and oils.
Nitrile	It is an excellent general duty glove material. Nitrile gloves provide protection from a wide variety of solvents, oils, petroleum products and some corrosives.
PVC	Provides excellent abrasion resistance and protection from most fats, acids, and petroleum hydrocarbons.
PVA	Gloves made of PVA are highly impermeable to gases. Provides excellent protection from aromatic and chlorinated solvents. Cannot be used in water or water-based solutions.
Viton	Provides exceptional resistance to chlorinated and aromatic solvents. Provides resistance to cuts and abrasions.
Silver shield	Resists a wide variety of toxic and hazardous chemicals. Provides the highest level of chemical resistance.
Natural rubber	Provides flexibility and resistance to a wide variety of acids, caustics, salts, detergents and alcohols.

Glove thickness, usually measured in mils or gauge, is another point of consideration. A 10 gauge glove is equivalent to 10 mils or 0.01 inches. Thinner, lighter gloves offer better touch sensitivity and flexibility, but may have shorter breakthrough times. Generally, doubling the thickness of the glove quadruples the breakthrough time. *Glove length* should be chosen based on the depth to which the arm will be immersed in or to the extent chemical splash is expected. The gloves longer than 14 inches provide extra protection against splash. *Glove size* also an important point of consideration. A particular sized glove (small, medium or large) may not fit to all the researchers in a laboratory. Gloves which are too tight tend to cause fatigue, while gloves which are too loose (Figure 19.2) will have loose finger ends which make laboratory work more difficult. The circumference of the hand, measured in inches, is roughly equivalent to the reported glove size. Glove colour, cuff design, and lining should also be considered for some tasks.

Glove Inspection, Use and Care

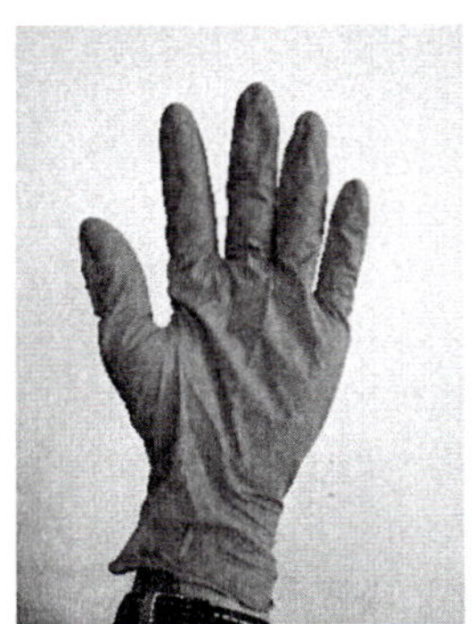

Fig. 19.2: Loose glove

Gloves should be inspected for signs of degradation or puncture before use. Test for pinholes can be made by blowing or trapping air inside and rolling them out. Do not fill them with water, as this makes the glove uncomfortable and may become more difficult in wearing it. Moreover, disposable gloves should be changed when there is any sign of contamination. Reusable gloves should be washed frequently if used for an extended period of time. While wearing gloves, be careful not to handle any other chemicals not involved in the procedure. Touching equipment, phones, wastebaskets, door handles or other surfaces may cause contamination. Wash hands immediately after removing the gloves with soap and water.

Removal of gloves after completion of work

Gloves should be removed carefully by avoiding contact with any exterior/contaminated parts of the. Disposable gloves should be removed as demonstrated in Figure 19.3.

- Grasp the exterior of one glove with other gloved hand.
- Carefully pull the glove off your hand, turning it inside-out. The contamination is now on the inner side.
- Slide your ungloved finger into the opening of the other glove. Avoid touching any exterior part of gloves.

- Carefully pull the glove off your hand, turning it inside out again. Thus, all the contaminants are contained.
- Discard the gloves appropriately.

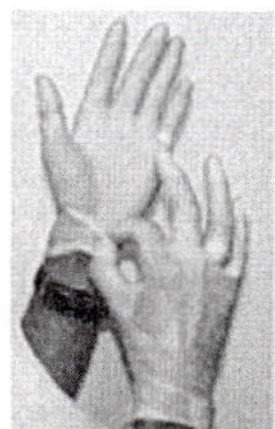 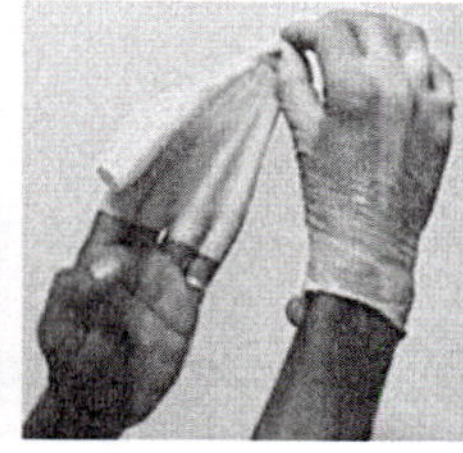 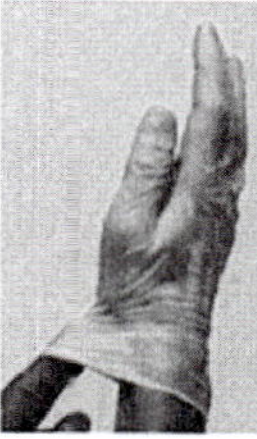 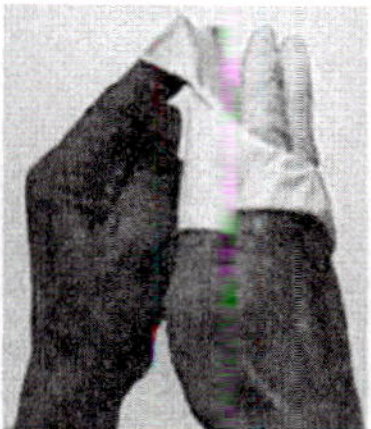 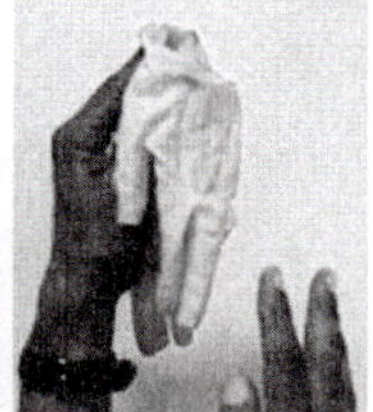

Fig. 19.3: Safe removal of gloves after use

Latex gloves and related allergies

Allergic reactions to natural rubber latex have increased since recommendations have been made for the use of gloves to protect against potentially infectious materials. Increased glove demand also resulted in higher levels of allergens due to changes in the manufacturing process. In addition to skin contact with the latex allergens, inhalation is another potential route of exposure. In addition to the latex proteins, the powders used to lubricate the interior of the glove might be responsible for allergic reactions. Following actions are suggested to reduce allergic reactions to latex:

- Whenever possible, substitute gloves made of another material.
- If latex gloves needs to be used, choose reduced-protein, powder-free latex gloves.
- Wash hands with mild soap and water after removing latex gloves.

Hearing Protection

Most of the laboratory equipment and operations do not produce noise levels that require the use of hearing protection. However, use of hearing protection, such as earplugs, earmuffs or hearing bands, may improve communication or provide comfort to the worker in a noisy laboratory environment. The most common noisy equipment (which may not appear to be much disconformable, but might be harmful) in a laboratory is sonicator. While using sonicator in a laboratory, the worker(s) should use hearing protection recommended by the manufacturer of the sonicator.

Eye Protection

Eyes are extremely vulnerable to biological infection, eye protector must be worn in the laboratory, particularly while using harmful chemicals. The eye protector includes:

- Goggles,
- Safety glasses,
- Face/splash shield.

For non-aerosol generating chemicals/activities, safety glasses and face shields provide adequate protection to eyes/face against splashes and spills. However, these do not provide substantial protection against airborne particles and aerosols. Goggles offer increased protection (by protecting the sides being somewhat air-tight) against spills and aerosol generating activities.

Foot Protection

Shoes used in a laboratory must be comfortable and should cover the entire foot. Don't wear sandals, high-heels, bathroom chappals or canvas shoes while working in a laboratory. Always prefer shoes made of leather or synthetic and fluid impermeable material. Closed-toed shoes should be worn at all times in a laboratory where chemicals are stored or used. Perforated shoes, sandals should not be worn in laboratories or where mechanical work is conducted. Such shoes offer no barrier between the foot of laboratory worker and chemicals or broken glass. Leather shoes tend to absorb chemicals and may have to be discarded if contaminated with a hazardous chemical/material.

Respiratory protection

Respiratory protection is generally not required while working with most BSL-1 and BSL-2 chemical/agents. However, respiratory protection is required from some BSL-2 agents when there is risk of aerosol generation which cannot be mitigated using alternative procedures or containment equipment. Respiratory protection can be provided by using either a mask or a respirator over the nose and mouth. Mask or respirator must not be reused and it is effective only when fits well to the worker's face. Head size, face size, and facial hair can affect the fit of the mask and reduce the efficacy to protect.

UV protection

The efficacy of UV light in biosafety cabinet (BSC) is debatable. Most studies indicate that the use of UV light is not an effective method of BSC decontamination. UV is harmful for eyes and skin; therefore, utmost care must be taken not to expose any naked part of the body to UV light. Life of a UV tube is limited; therefore, do not keep it on for unnecessarily longer period. In most of the cases, 10-15 minutes of exposure is more than sufficient for surface decontamination.

2. General Laboratory Safety Rules

The following rules should be included in laboratory safety policies and must be followed in every laboratory. These include what one should know in the event of an emergency, proper signage, safety equipment, safe use of laboratory equipment, and some of the basic rules (Figure 19.4).

1. Be sure to read all fire alarm and safety signs and follow the instructions in the event of an accident or emergency.
2. Ensure you are fully aware of your facility/building evacuation procedures.
3. Make sure you know where your laboratory safety equipment (including first aid kit, fire extinguishers, eye wash stations and safety showers) is located and how to properly use them.
4. Know emergency phone numbers to call for help in the case of an emergency.
5. Laboratory areas containing carcinogens, radioisotopes, biohazards, and laser emitters should be properly marked with the appropriate warning signs.
6. Open flames should never be used in a laboratory unless it is permitted.
7. Make sure you are aware about fire alarms located in laboratory.
8. Always turn off instruments after use (when not in use).
9. When there is a fire drill, be sure to turn off all the electrical equipment and close all containers.
10. Always work in properly ventilated areas.
11. Do not chew gum, drink, or eat while working in a laboratory.
12. Laboratory glassware should never be utilized as food or beverage containers.
13. Each time using glassware, be sure to check it for chipping off and cracks. Notify your lab supervisor of any damaged glassware, so that it can be properly disposed of/replaced.

14. Never use lab equipment yourselves for which you have not been trained.
15. If an instrument is not operating properly, report the issue to a technician right away. Never try to repair it on your own.
16. If you are the last person to leave the lab, make sure to switch-off the electrical equipment not in active use, close water taps, lock all the doors, turn off ignition sources, etc.
17. If there is no urgency, then do not work alone in a laboratory.
18. Never leave an on-going experiment unattended.
19. Never lift any glassware, solutions, or other types of apparatus above eye level.
20. Do not pipette using mouth sucking.
21. Always follow appropriate procedures for disposing laboratory waste.
22. Report any injuries, accidents, and broken equipment or glass right away, even if the incident seems small or unimportant.
23. If you have been injured, tell out immediately and as loud as you can to ensure you get necessary help quickly.
24. In the event of a chemical splashing into your eye(s) or on your skin, immediately flush the affected area(s) with running water for at least 20 minutes.
25. If you notice any unsafe conditions in a laboratory, let your supervisor know it as soon as possible.
26. Don't taste or sniff chemicals present in any form in a laboratory, as it can be highly dangerous and even deadly. The most appropriate way to know what is exactly present in the beaker/glass bottles is to label it properly.
27. Follow all the safety guidelines made for using radioactive materials in the laboratory.
28. Always discard agarose/polyacrylamide gel following appropriate procedure. Don't discard them in general/regular dustbin.
29. Discard contaminated/used culture media appropriately after autoclaving
30. Discard culture media plates after autoclaving to prevent them from contaminating soil and environment.

Fig. 19.4: Housekeeping rules to be followed while working in the laboratory

3. Dress Code Related Safety

Laboratory dress codes set a policy for the employees to avoid/prevent accidents in the laboratory. For example, shorter clothing might look nice or enjoying the warm weather outside, but it increases chances of accidental contact with dangerous chemicals in laboratory. Following dress/safety rules can be followed for working in a laboratory:

- Always tie back hair that is chin-length or longer
- Make sure that loose clothing or dangling jewellery is not appropriate for a laboratory worker.
- Never wear sandals or other open-toed shoes in a laboratory. Footwear should always cover the foot completely.
- While working with Bunsen burner, lighted splints, matches, etc. should not be allowed.

B. Chemical Safety

Detailed safety procedures for chemical safety have been already discussed in chapter 1; however, following are the general guidelines for safe handling of chemicals in a laboratory.

1. Always label the containers containing chemical.
2. Use protective equipment for protection and wear a laboratory coat.
3. Avoid smelling, inhaling, and tasting a chemical.
4. Avoid direct contact with chemical to your hands face, clothes, and shoes.
5. Hazardous chemical must be used only as directed.
6. Attach necessary information about the chemical on the container.
7. Read the warning label while opening a newly received reagent/chemical. This will not only help proper storage but also from hazards from the chemical.
8. While leaving the laboratory, wash your hands thoroughly with soap and water.
9. Wear necessary personal protective equipment (e.g., eye protection, laboratory coats, gloves, and face protection) in the laboratory.
10. Work carefully with sharp objects (such as needles, scalpels, Pasteur pipettes, and capillary tubes) in the laboratory.
11. Work carefully to minimize potential formation of aerosol. Confine aerosols as close as possible to their source of generation by using appropriate biosafety cabinet.

C. Waste Disposal and Cleanliness

Cleaner laboratory/glassware/chemical is essential in chemistry. Tolerance level for contaminants varies with the experiment/work you perform, and sometimes a chemist does not know how much important is cleanliness of glassware/chemical for an experiment until it fails.

General Cleaning Tips

- Disassemble your apparatus as soon as possible after you are finished with it. Rinse appropriate parts with suitable solvent.

- Graduated cylinders, beakers, Erlenmeyer flasks, burettes, pipettes, funnel, etc. used to dispense reagents need to be cleaned with tap water and then rinsed with distilled/deionised water, stored/air dried on a rack.

General Cleaning Procedure

Following steps should be used for cleaning the glassware for which a simple solvent rinse is not sufficient.

- Degrease the ground glass joint of glassware by wiping them with a paper towel soaked in ether, acetone or other appropriate solvent.
- Wear appropriate gloves and minimize exposing yourselves to the vapour/aerosol.
- Place the glassware in a warm concentrated aqueous solution of detergent, and let it dipped in for several minutes.
- Scrub with brush, and rinse thoroughly with tap water followed by a final rinse with deionised water.
- If greasy residue is still present on the glass surface, more aggressive action must be taken (as mention in Chapter 3).

Waste Disposal

Wastes are unwanted/unusable materials; these are discarded after use. All the laboratory staff should take active participation in proper disposable of the laboratory wastes.

Basic Norms to Dispose Laboratory Waste

- Do not store waste in a metal container.
- Do not store the chemical waste under a fume hood.
- Properly label the containers of different wastes.
- Store waste in a container with lid and does not get punctured with sharp waste material.
- Do not dispose-off incompatible chemical wastes in a same container.

Type of Waste

- Normal municipal waste (general)
- Recyclable waste

- Broken glass
- Biomedical/medical waste
- Chemical waste
- Sharps, broken glass
- Radioactive waste
- Electronic and computer waste.

Laboratory Waste can be Divided into two Categories

Can be disposed into sink	Cannot be disposed into sink
• Solution with pH 5 – 9. • Solution with low concentration, e.g., weak acid, weak alkali, etc.	• Solid waste. • Solution with a pH >9 and <5. • Organic solvent. • Toxic substance. • Heavy metal. • Organic waste, e.g. grease, oil, oil paint, hydrogen peroxide, microbes, carcasses.

Disposal of Laboratory Waste

Disposal of sharps

Sharp wastes include clinical/laboratory needles, blades, laboratory broken glasses capable of causing punctures or cuts. Therefore, to avoid injuries sharp wastes must be handled carefully following the below mentioned steps.

1. Collect all sharps (including needles) in approved plastic sharps containers.
2. Chemically decontaminate the infectious items prior to disposal into the container. Once the container is filled, it can be autoclaved.
3. Neutralize the chemical decontaminants using neutralization solution.
4. Securely close/cover the container.
5. Place the filled container to a designated area for pick-up and disposal.
6. Do not put sharps into plastic bags.

Decontamination of Solution Containing Ethidium Bromide

Ethidium bromide (EtBr) is a powerful mutagen and is moderately toxic also. Gloves must be worn while working with the solutions containing EtBr. After use, these solutions must be decontaminated following one of the methods described below as per the need.

Decontamination of concentrated solution (containing >0.5 mg/ml) of EtBr

Method–1

This method minimises mutagenic activity of ethidium bromide (in a Salmonella/microsome assay) by approximately 3000-fold. The procedure include following steps.

1. Add sufficient water to reduce the concentration of EtBr to <0.5 mg/ml.
2. Add equal volume of 0.5 M $KMnO_4$, mix carefully, and then add 1 volume of 2.5 N HCl. Mix properly with care, and allow the solution to stand at room temperature for six hours.

   Caution: $KMnO_4$ is an irritant and explosive. Solutions containing $KMnO_4$ should be handled in a chemical hood. Concentrated HCl (commercially available as 5N) must be handled carefully. Extra care is required while diluting conc. HCl (add acid into water and not *vice versa*) and adding HCl into the $KMnO_4$ solution.
3. Finally, add equal volume of 2.5 N NaOH, mix carefully, and the decontaminated solution can be discarded now.

Method–2

This method reduces mutagenic activity of ethidium bromide (in Salmonella/ microsome assay) by approximately 200-1000 fold.

1. Add sufficient water to reduce the concentration of ethidium bromide to <0.5 mg/ml.
2. To this diluted solution, add 0.2 volume of fresh 5% hypophosphorus acid and 0.12 volume of fresh 0.5 M sodium nitrite. Mix carefully and leave the mixture for 24 hours at room temperature.

   Important: Check that pH of the solution is < 3.0 (use pH paper or strip, not pH meter).

   Note: Hypophosphorus acid is usually supplied as a 50% solution, which is corrosive and should be handled with care. It should be freshly diluted immediately before use. Sodium nitrite solution (0.5 M) should be freshly prepared by dissolving 34.5 g of sodium nitrite in water to a final volume of 500 ml.
3. After incubation for 24 hours at room temperature, add an excess of 1 M sodium bicarbonate. The decontaminated solution can be discarded after this treatment.

Decontamination of less concentrated solution (containing ≤0.5 mg/ml) of EtBr (e.g. electrophoresis buffer)

Method-1

1. Add 2.9 g of Amberlite XAD-16 for each 100 ml of solution. Amberlite XAD-16, a nonionic, polymeric absorbent.
2. Store the solution for 12 hours at room temperature with shaking it intermittently.
3. Filter the solution through a Whatman No.1 filter, and discard the filtrate.
4. Seal the filter and Amberlite resin in a plastic bag, and dispose of the bag in hazardous waste.

Method-2

1. Add 100 mg of powdered activated charcoal for every 100 ml of EtBr containing solution.
2. Store the solution for 1 hour at room temperature with shaking it intermittently.
3. Filter the solution through a Whatman No.1 filter and discard the filtrate.
4. Seal the filter and activated charcoal in a plastic bag, and dispose of the bag in hazardous waste.

Decontamination of Surfaces Contaminated with Ethidium Bromide (e.g. spill of solution containing < 0.5 mg/ml EtBr)

In case of spills of EtBr solution on surfaces in the work area, it should be decontaminated immediately following the method described below:

1. Spread slurries of Amberlite XAD-16 or activated charcoal powder over the spill area so as to soak the liquid completely.
2. Collect the slurry/powder carefully and put in a polythene bag for disposal.
3. Wipe the surface with ethanol. Use gloves during the decontamination process.

Notes

i. Treatment of dilute solutions of EtBr with hypochlorite (bleach) is not recommended as a method of decontamination. Such treatment reduces the mutagenic activity of EtBr (in the Salmonella/microsome assay) by about 1000-fold, but it converts the dye into a compound that is mutagenic in the absence of microsomes.

ii. Ethidium bromide decomposes at 262 °C and is unlikely to be hazardous after incineration under standard conditions.

Let's Check Your Understanding

1. What is the main purpose of Personal Protective Equipment (PPE) in a laboratory?
   a. To reduce or eliminate hazards
   b. To protect the wearer from exposure to hazardous materials
   c. To improve the aesthetic of the laboratory
   d. To reduce the cost of laboratory operations.
2. Which of the following is NOT considered a type of disposable protective clothing?
   a. Laboratory coats
   b. Hair and shoe covers
   c. Synthetic over-garments
   d. Protective sleeves.
3. What should be considered when selecting gloves for laboratory use?
   a. Colour of the gloves
   b. Cost of the gloves
   c. Length of cuff, thickness of material, and allergies of laboratory personnel
   d. Manufacturer's popularity.
4. What is the recommended action if a spill occurs or a breach in the glove material is discovered?
   a. Continue working and change gloves later
   b. Change gloves immediately
   c. Wash the gloves and reuse them
   d. Inform a supervisor and continue working.
5. What type of laboratory footwear is recommended for safety?
   a. Sandals
   b. High-heels
   c. Closed-toed shoes made of leather or synthetic fluid impermeable material
   d. Canvas shoes.

6. Which type of glove offers the highest resistance to permeation by most gases and water vapour?

   a. Nitrile
   b. Butyl
   c. PVC
   d. Viton.

7. What is the primary concern while using UV light in a biosafety cabinet?

   a. It is highly effective for surface decontamination
   b. It can be harmful to eyes and skin
   c. It is cost-effective
   d. It is easy to use and maintain.

8. What should you do immediately after a chemical splashes into your eye(s) or on your skin?

   a. Wait for help to arrive
   b. Flush the affected area(s) with running tap water for at least 20 minutes
   c. Apply an ointment
   d. Rub the affected area gently.

9. What is the proper method for disposing of broken glass in a laboratory?

   a. Throw it in the general trash bin
   b. Place it in a designated sharps container
   c. Store it under a fume hood
   d. Dispose it with chemical waste.

10. How should solutions containing Ethidium Bromide (EtBr) be decontaminated if the concentration is greater than 0.5 mg/ml?

    a. Pour it down the sink
    b. Use bleach and water
    c. Follow either of the Methods described for high concentration solutions
    d. Store it in a metal container for later disposal.

Answers

1. b
2. a
3. c
4. b
5. c
6. b
7. b
8. b
9. b
10. c

20

Biosafety and Ethical Issues of Laboratory Research

Introduction

Several hazardous chemicals, microbes, and techniques like recombinant DNA technology are used for research and development work in research Institutes and universities. Therefore, the researchers, technical, helping and supporting staffs as well as the laboratory outputs must be kept free from contaminations. Thus, biosafety is necessary to prevent unintentional exposure/ release of hazardous chemicals, microbes/pathogens, and toxins to the laboratory staff and environment. Following the biosafety guidelines and ethical considerations are essential to maintain safety standards for the researchers as well as the environment. This chapter deals with various biosafety measures and ethical issues to be strictly followed while working in a research laboratory. The required safety level needs to be determined as per the chemical/biological materials are used in the laboratory to keep the researcher/environment safe for sustainable research and developmental activities.

Biosafety

The term biosafety is used to describe the procedure and policies needed to ensure environmental and personal safety. Biosafety refers to the containment principles, technologies and practices that are implemented to prevent unintentional exposure to pathogens, toxins, or their accidental release into the environment. A fundamental objective of any biosafety program is the containment of potentially harmful chemical/biological agents as precautionary safety measures.

Why Biosafety?

With the increasing number of countries adopting molecular biology tools and techniques in life science research and development activities, especially in the areas of agriculture and medicine, the biosafety issues are gaining importance to ensure biological safety for the public and the environment. Conducting research in safe and sustainable manner is not only a personal requirement but essential collective efforts to ensure biological safety for a clean and safe environment. This requires rules, regulations, monitoring bodies, and awareness among the workers as well as the public. Therefore, biosafety *per se* is an integral part of laboratory research and requires awareness among the researchers so that biological safety can be well taken care from the grassroot level.

Biosafety of Genetically Modified Organism (GMO)

GMOs are now getting commercial applications in agriculture, healthcare, industries, etc. often for better quality of the products. However, there are key differences in these sectors with respect to the regulatory point of views. Healthcare industry is highly regulated and the products are generally life-saving drugs where certain minor risks can be easily compromised with the life-saving benefits. The key players of the healthcare industry (medical practitioner, drug industry and the regulatory authorities) are often well aware of the biosafety concerns. The GMOs and the purified products are handled under contained environment. There is minimized public concern associated with the use of GMOs in healthcare industry. On the other hand, agriculture deals with the crop and animal husbandry; protecting them from insect pests and diseases; improving their taste, quality and acceptability to the consumers, as well as studying their nutritional and associated effects. Moreover, agricultural operations are mainly performed by rural farmers having little awareness/understanding about biosafety requirements and measures. With the adoption of GM crops in 29 countries, grown over 190 million hectares by 17 million farmers world over, the concerns associated with GMOs are widely discussed. However, the concerns differ greatly, depending on the particular gene–organism combination. Therefore, a case-by-case approach would be required for the assessment of associated biosafety concerns. The safety concerns in agriculture not necessarily associated with the characteristics of the products but the way it is produced, particularly in case of human food. European Union has been the opponent of the GM technology based on the associated biosafety concerns. However, the studies conducted during 2011 in Europe confirmed that GM crops are safe as animal feed. Since GM crops and animals are grown under open environment and they can interact with other organisms in the surrounding environment, GMOs in agriculture have become a more sensitive issue than they are in the healthcare industry. The potential risks of growing GMOs and using their products may be associated with (i) human health, (ii) environmental concerns, and (iii) social and ethical issues. The potential risks from the use of GMOs and their products broadly fall under three categories, i.e. the risk to human health, environmental concerns, and social & ethical issues.

Risk to Human Health

Risks of GMO to human health are related mainly to toxicity, allergenicity and antibiotic resistance of the organism/product.

Risk of Toxicity

The risk of toxicity may be directly related to the nature of the product whose synthesis is controlled by the transgene or the changes in the metabolism and the composition of the organism resulting from gene transfer. For example, in some cases organism may contain inactive pathways leading to the production of toxic substances. Addition of new genetic material could reactivate these inactive pathways or increase the level of toxic substances within the plant/ organism. Most of such toxicity risks can be assessed using scientific methods, both qualitatively and quantitatively.

Risk of Allergy

Production of GMO generally includes the introduction of newer proteins from the organism which has not been consumed as food. There is a risk of such proteins becoming allergens since virtually all known food allergens are proteins. The possibility of transferring allergens with genetic engineering came to light when a methionine producing gene from the Brazil nut was incorporated into soybean to enhance its nutrient content. The process was experimented by Pioneer Hi-bred in the United States. The tests conducted on allergens confirmed that consumption of the transgenic soybean could trigger an allergic response in those sensitive subjects who were allergic to Brazil nut. The company, therefore, decided not to release the transgenic soybean for sale.

Antibiotic Resistance

Production of GMO also involves use of genes that provide antibiotic resistance as selectable markers. Early in the modification process in the laboratory, these selectable markers help in selecting cells that have taken up the foreign genes. Use of these antibiotic resistance markers has raised the concerns that eating food materials carrying antibiotic resistance marker would reduce the effectiveness of antibiotic used as medicine to fight microorganisms/diseases. The antibiotic resistance gene produces enzyme that can degrade antibiotic. Therefore, theoretically if a transgenic tomato with an antibiotic resistance gene is eaten at the same time as an antibiotic; it could destroy the antibiotic in the stomach.

Eating Foreign DNA

There have been apprehensions about eating foreign DNA through GM foods (i.e. the pieces of DNA that did not originally occur in the food plant). DNA being present in all living things, such as plants, animals, microorganisms, is eaten by human beings with every meal. Most of it is broken down into their basic molecules during the digestion process, whereas a small amount that is not broken down is either absorbed into the blood stream or excreted. So far there is no evidence that DNA from GMO including transgenic crops is more dangerous to human health than DNA from conventional crops, animals or associated microorganisms that are normally eaten. According to FAO/WHO documents on safety aspects of GMOs of plant origin, the amount of DNA which is ingested varies widely, but it is estimated to be just 0.1 to 1.0 gram per day. Novel DNA from a GM crop would represent less than 1/250,000 of the total amount consumed. This means that the possibility of the transfer of a gene that have been introduced through genetic modification is extremely low. In fact, it would require that all of the following events would occur:

- The relevant gene in the plant DNA would have to be released, probably as an intact linear fragment;
- The gene would have to survive harvesting, preparation and cooking processes, in addition to degradation by nucleases;
- The gene would have to compete for uptake with other dietary DNA;
- The recipient bacteria or mammalian cells would have to be able to take up the DNA/gene and it will have to survive enzymatic digestion;
- The gene would have to be inserted into the person's genome by very rare recombinant events;
- The gene would have to be stably integrated and expressed in the new host.

Changes in Nutritional Level

There have been concerns about the accidental changes in the nutritional components of transgenic crops while incorporating other traits. For example, the comparison of isoflavone level in the Roundup Ready soybeans (GM variety incorporated with herbicide tolerant gene) with the conventional varieties showed a minor difference. Because isoflavones are thought to play a role in preventing heart disease, breast cancer, and osteoporosis, this became a matter of concern and it was investigated seriously. However, studies in support of applications for permission to sell transgenic soybean indicated similarity in commonly tested nutritional components in transgenic and conventional variety.

Risk to Environment

Introduction of GMO in the environment necessitates a close examination for potential ecosystem disruption. Ecological risks include the impact of introduced trait introgressed into other related species due to out crossing, the potential build-up of resistance in insect populations to engineered insecticidal traits, unintended secondary effects on non-target organisms, and potential effects on biodiversity.

Persistence of Transgene or Transgene Product

The gene transferred into an organism or the resultant products can actually remain in environment leading to environmental problems. For example, in case of Bt crops it was suspected that insecticidal proteins can persist in the environment, but experiments have proved that these are degraded in the soil. There are also concerns in case of microorganisms about their capacity to adapt to new environmental conditions and persist in the environment as spores.

Interaction with Non-target Organism

The intentional release of GMO into the environment has led to an increased interest in possible interactions that may occur between other organisms in the environment.

Unpredictable Gene Expression or Transgene Instability

Unintended genomic changes may occur as a secondary consequence of genetic modification. Such changes can lead to production of new proteins that may be toxic or allergenic or may disrupt or alter metabolic pathways that play a role in making the GMO successful.

Gene Flow

Accidental cross breeding between GMO (plant) and traditional varieties through pollen transfer can contaminate the traditional local varieties with the transgene resulting in the loss of traditional varieties. The wind, rain and insect pollinators can contribute to the spreading of pollens resulting in the contamination of local varieties through cross pollination/breeding. One such example, which has been reported, is that of Mexican corn. Recently it has been reported that wild corn variety located in some remote areas of Mexico have been contaminated by gene from GMO. The consequences associated with such gene transfers may impact intellectual property, increase weediness if transfer to compatible weedy relatives, or lead to extinction of endangered varieties of the genera.

Increased Weediness

Weediness means the tendency of a plant to spread beyond the field where it was first planted. For example, "*Jal kumbhi*" was introduced into India as an ornamental plant, but now it has become a major invasive weed in the water-logged areas. There are apprehensions about GM crops becoming weeds. For example, a salt tolerant GM crop if escapes into marine areas could become a potent weed there. There is also fear about the development of superweeds (i.e. a weed that has acquired the herbicide tolerant gene due to genetic contamination with a herbicide tolerance GMO through in-field cross-breeding to related species or through horizontal gene transfer), which cannot be controlled by the herbicide.

Loss of Biodiversity

There have been concerns about reduction in the genetic diversity in cropping systems by the development and global spread of improved crop varieties. This genetic erosion has occurred as the farmers have replaced using traditional varieties due to monoculture. This is expected to further intensify as more and more transgenic crops are introduced which bring in considerable economic benefits to the farmers.

Changes in Soil Ecology

Many plants leak chemical compounds into the soil through their roots. There are concerns that transgenic plants may leak different compounds compared to that by conventional plants. Speculations are that this may change the ecology of the soil in terms of functional composition and biodiversity.

Important Aspects of Biosafety

Though there has been increasing awareness about biosafety all over the globe, there are still lots to be understood and followed in laboratory research, particularly at grassroot level in the developing countries. Many countries have put into place effective regulatory procedures that are much more rigorous for GM than non-GM foods. However, formulation of basic guidelines with multilevel regulatory bodies is required for compliance of biosafety measures. Though biosafety regulatory bodies may exist, their effective functioning is a big question in many of the countries using GMO and their products. The need of the day is to strengthen regulatory bodies and create awareness among the research mangers, scientists and students about the biosafety requirements. The internal regulatory body at research institute level includes an Institutional Biosafety Committee (IBSC) or its equivalent body consisting of experts from different relevant disciplines. The IBSC must ensure (i) existence of the basic biosafety equipment required according the safety level of the experiments conducted, organize trainings for the researchers to make new comers aware of the biosafety measures as well as to update their knowledge of biosafety.

As the ultimate aim is to eliminate or minimize biological contamination, there are three important aspects of biosafety and biosecurity:

1. ***Biological hazard:*** is the potential risk of uncontrolled exposure to biological agents that cause disease.

2. ***Biocontainment:*** includes the measure used to prevent release/spread of infectious microorganisms from research centers or other places where they may be produced/stored.

3. ***Bioprotection:*** is a set of measures taken to reduce the risk of loss, theft, misuse or intentional release of pathogens and toxins, including those governing access to facilities, materials storage and data and publication policies.

Biosafety is a complex discipline which is not devoid of dangers. That is why it is so crucial to have a set of rules and barriers that prevent biological hazards derived from exposure to infectious biological agents. The following principal elements should be maintained in order to ensure biosafety in laboratories:

1. ***Standards:*** Individuals (students, lab attendant, researchers including SRFs and students) who handle potentially contaminated biological agents must be aware of the risks and master the practices and techniques required to do their jobs with safely.

2. ***Universality:*** Biosafety measures must be observed by everyone who enters in laboratories or even in surrounding areas, because everyone is at risk of carrying pathogenic microorganisms.

3. ***Barriers:*** The elements used to contain biological contamination can be of different types including (i) immunization (vaccines), (ii) primary barriers like safety equipment including gloves, coat, masks, and (iii) secondary barriers starting from insulation of work areas to use of hand washers and ventilation systems.

4. ***Elimination:*** All the waste materials generated must be disposed off with strict compliance as per the specific procedure suited to the type of waste.

Biosafety Level in Laboratory Research

Biosafety level (BSL) are used to identify the protective measures needed in a laboratory setting to protect workers, the environment, and the public. There are many ways to combine equipment, practices, and laboratory design/ features to achieve appropriate biosafety and biocontainment. These are determined through biological risk assessments specifically conducted for each experimental protocol. According to WHO Safety Manual, there are four levels of biosafety, these are:

Biosafety Level 1 (BSL-1)

BSL-1 labs are used to study infectious agents or toxins not known to consistently cause disease in healthy adults. This requires following basic safety procedures called Standard Microbiological Practices and does not require special equipment or laboratory design/feature.

Biosafety Level 2 (BSL-2)

BSL-2 laboratories are used to study moderate-risk infectious agents/toxins that pose a risk if accidentally inhaled, swallowed, or exposed to skin. Laboratory design requirements include hand-wash sink, eye washing station, and doors that close automatically and lock. The labs must also have access to equipment that can decontaminate laboratory waste, including an incinerator, an autoclave, and/or methods depending on the biological risks involved.

Biosafety Level 3 (BSL-3)

Laboratories that are used for research on infectious agents/toxins that may be transmitted through air and cause potentially lethal infection through inhalation exposure require BSL-3. Researchers perform experiments in biosafety cabinets that use carefully controlled air flow or sealed enclosures to contain infectious microorganisms. The laboratories are designed to be easily decontaminated.

Biosafety Level 4 (BSL-4)

Laboratories that are used to study infectious agents or toxins that pose a high risk of aerosol-transmitted laboratory infections and life-threatening infection/disease for which no vaccine or therapy is available require BSL-

4. The laboratories incorporate all Biosafety Level 3 features and occupy safe, isolated zones within a larger building or may be housed in a separate, dedicated building. Access to these laboratories is carefully controlled and requires significant training.

Risk Assessment

Risk assessments are conducted by evaluating the way in which the infectious agent/toxin is transmitted and its ability to cause infection/health hazard to the worker, the activities performed in the laboratory, the safety equipment and design elements present in the laboratory, the availability of preventive medical countermeasures or treatment, and the health and training of the laboratory worker.

One of the most helpful tools available for performing a microbiological risk assessment is the listing of risk group for microbiological agents.

- ***Risk group 1***: No or low individual/community risk. A microorganism that is unlikely to cause human or animal disease.
- ***Risk group 2***: Moderate individual risk, low community risk. A pathogen that can cause human or animal disease but unlikely to cause serious hazard to laboratory workers, the community, livestock or the environment. Laboratory exposure may cause serious infection, but effective treatment.
- ***Risk* group 3**: High individual risk, low community risk. A pathogen that usually causes serious human or animal disease but does not ordinarily spread from one infected individual to another. Effective treatment and preventive measures are available.
- ***Risk group 4***: High individual and community risk. A pathogen that usually causes serious human or animal disease and that can be readily transmitted from one individual to another, directly or indirectly. Effective treatment and preventive measures are not usually available.

Ethical Issues in Laboratory

In recent time of modern scientific era, ethical concerns are making place everywhere in almost all sectors. Several ethical issues exhibit its effect within the laboratory techniques and during the working procedures. The foremost ethical challenges such as consent, confidentiality, proficiency, codes of conduct, conflict of interest, sharing and access to laboratory are some of the issues becoming more widespread in resource-limited settings of laboratory. Irrespective of the laboratory results, laboratory staff/worker is another vital component of ethical issues. Hence, it is important to understand the ethical issues to help improve confidence, operational reliability, capability, neutrality, and security of the staff/workers. There are several organizations, such as International Organization for Standardization (ISO), International Federation of Clinical Chemistry (IFCC), and American Association of Clinical Chemistry (AACC), that have defined ethical recommendations for different operations in laboratories.

There are numerous ethical issues in laboratory, which can be grouped into three phases according to the laboratory work allocation.

1. Pre-analytical phase,
2. Analytical phase,
3. Post-analytical phase.

Pre-analytical Phase

Herein, the issues are mainly concerns with the collection of samples, its handling and safe storage in laboratory. Any information regarding demography of samples, testing in laboratory, tests that have been ordered, and requisition of the test should be shared only with appropriate personnel. Collection of samples should be done as per universally suggested safety measures so as to protect the originality of the sample.

Analytical Phase

In a laboratory setting, privacy, excellence, and proficiency are very crucial. Therefore, lab personnel should do their work in a proper manner. One needs to behave properly with other staff during sample collection, analysis as well as appropriate attitude towards handling of sophisticated lab instruments, which are some of the ethical concerns in a laboratory.

Post-analytical Phase

This includes reporting and elucidation of the results, storage of remaining sample for future usage, and access to the data. While giving final result, one should maintain purity/genuinely of the work without any manipulation in the data. While publishing any work from the laboratory, it must include the name of all the people who have contributed in the work giving time and efforts. Any gift authorship should be discouraged and one should maintain sanitation of the work place to make pleasant atmosphere in the laboratory.

Some of the ethical issues of laboratory include:

- **No maleficence** relates with not causing harm to self or other personnel in the laboratory,
- **Beneficence** relates with undue favour to anybody inside or outside the laboratory,
- **Autonomy** relates with making free and informed choice,
- **Justice** is associated with treating people fairly, equally, and without any bias,
- **Fidelity** relates with keeping promises and agreement,
- **Honesty** is needed to avoid any defraud, deceive or misleading act,
- **Privacy** is needed to respect personal privacy and confidentiality,
- Fabrication, falsification, and plagiarism are needed to avoid any manipulation or mis-presentation of experimental/literature data,
- Social responsibility is needed to avoid causing any harm to society and working for social benefits,
- Legality necessitates to obey laws and institutional regulations concerning the research work,
- Opportunity relates with providing fair opportunities to all with regard to use of the resources.

Further Readings

Kumar S (2012). Biosafety Issues in Laboratory Research. Biosafety 1: e116. doi:10.4172/2167-0331.1000e116

Kumar S (2014). Biosafety Issues of Genetically Modified Organisms. Biosafety 3: e150. doi:10.4172/2167-0331.1000e150

Kumar S (2015). Biosafety and Biosecurity Issues in Biotechnology Research. Biosafety 4: e153. doi:10.4172/2167-0331.1000e153

Let's Check Your Understanding

1. Which one of the followings is not an element of biosafety principle?

   A. Barriers
   B. Universality
   C. Truthfulness
   D. Elimination

2. A pathogen that usually causes serious human or animal disease and that can be readily transmitted from one individual to another, directly or indirectly. The pathogen should be classified under which of the following risk groups?

   A. Risk Group 1
   B. Risk Group 2
   C. Risk Group 3
   D. Risk Group 4

3. laboratory is used to study infectious agents or toxins that may be transmitted through the air and cause potentially lethal infection through inhalation exposure?

   A. BSL-1
   B. BSL-2
   C. BSL-3
   D. BSL-4

4. Ethical issues in laboratory can be grouped into how many phases as per the laboratory work allocation.

5. High individual risk, low community risk belongs to which risk group of biosafety?

6. Biosafety measures are essential to prevent unintentional exposure to hazardous chemicals, microbes, pathogens, and toxins to the laboratory staff and environment.

   True / False

7. The healthcare industry is less regulated than the agriculture sector regarding the use of genetically modified organisms (GMOs).

   True / False

8. The risk of toxicity from GMOs is primarily related to the nature of the product controlled by the transgene or changes in the metabolism and composition of the organism resulting from gene transfer.

   True / False

9. Laboratories that are used to study infectious agents or toxins that pose a high risk of aerosol-transmitted laboratory infections and life-threatening diseases require Biosafety Level 2 (BSL-2).

   True / False

10. Ethical issues in laboratory settings only concern the analytical phase of laboratory work.

    True / False

Answers

1. C
2. D
3. D
4. Three (Pre-analytical phase, Analytical phase and Post-analytical phase)
5. Risk group 3
6. True
7. False
8. True
9. False
10. False

21

Do's and Don'ts in Laboratory

~Precaution is better than cure!

A laboratory is essential for research work in science and development. Most of the research institutes have dedicated laboratories for researchers and students to conduct experiments, perform research work, as well as to provide hands-on learning/training to students. There are certain rules and regulations laid down by the laboratory in-charge which need to be followed strictly for safety purposes. Some of the laboratories keep 'safety manual' for researchers to read before entering into the laboratory to conduct experiments. Whenever safety is compromised in a laboratory, there is increase in risk of unwanted incidents in the laboratory. Storing food/beverages in the same refrigerator where laboratory chemicals are kept may lead to disastrous effects on health. This chapter aims to provide instructions in brief on the things to be performed (Do's) before and after entering onto a laboratory as well as the things never to be done (Don'ts) in a laboratory for the safety reasons

Good Laboratory Practices (Do's)

- ✓ Wear apron/labcoat while working in the laboratory.
- ✓ Use gloves/respiratory mask/fumehood while handling toxic/volatile/inflammable chemicals/reagents.
- ✓ Use pippet or pippetor for safety in transfer of solutions.
- ✓ Protect eyes, face and naked parts of the body from UV radiations. Use safety goggles or face mask/apron.
- ✓ Keep your eyes away from Liquid Nitrogen and use cryogloves while handling it.
- ✓ Arrange all the necessary materials/solutions before starting the experiment.
- ✓ Periodically fumigate/ozonize culture and inoculation rooms.
- ✓ Autoclave all contaminated materials before discarding.
- ✓ Pre-treat gel or solution containing Ethidium Bromide (EtBr) before discarding into laboratory waste-disposal bin, and use gloves while handling them.
- ✓ Burn all the transgenic plant materials in an incinerator after taking necessary observations.
- ✓ Make entry in the designated log-book and clean the instrument after using it properly.
- ✓ Switch-off the instrument after use or when not in use.
- ✓ Wash hands with soap and remove the labcoat before leaving/going out of the laboratory7.
- ✓ Ensure that switches for fans/lights, and water taps are off while leaving the laboratory.
- ✓ Follow the biosafety guidelines suggested for handling and disposal of the hazardous chemical.

Good Laboratory Practices (Don'ts)

- X Avoid eating, drinking and smoking in the laboratory.
- X Don't practice mouth pipetting of chemicals in the lab.
- X Never inhale, breath or smell-test chemicals in lab.
- X Don't leave infectious materials unattended in lab.
- X Don't blow out infectious material out of the pipette.
- X Never touch Ethidium Bromide (EtBr) containing gel/solution with necked hands.
- X Never discard EtBr containing materials in drain or wash-basin without appropriate pretreatment.
- X Don't reuse the plasticware that has been/to be used for PCR, rDNA and transformation work.
- X Don't touch Liquid Nitrogen; this may cause cold-burning. Keep your eyes away from its vapour.
- X Don't keep the door of a deep-freezer or refrigerator open for a longer period.
- X Don't try to open lid of a centrifuge before rotor stops.
- X Don't open lid of an autoclave until pressure releases completely, protect your hands and face from the out- coming hot steam.
- X Avoid overloading an electrical point or extension board in the laboratory.
- X Never use water to extinguish electrical fire.

Let's Check your Understanding

1. It is acceptable to store food materials and beverages in the same refrigerator along with laboratory chemicals?

   True / False

2. Reading the safety manual present the laboratory is recommended before starting/conducting any experiment in the lab.

   True / False

3. Wearing appropriate personal protective equipment (PPE) such as lab-coat, gloves, and/or goggles, as it might be necessary, is mandatory for working in a laboratory.

   True / False

4. It is safe to wear open-toed shoes or sandals in a laboratory.

   True / False

5. Always wash your hands thoroughly with soap and water after using any laboratory chemical.

   True / False

6. Mouth pipetting is a safe and acceptable practice in a laboratory.

   True / False

7. Leaving an on-going experiment unattended in a laboratory is permissible.

   True / False

8. Always ensure that chemicals and reagents are properly labelled and stored in the designated areas.

   True / False

9. Eating, drinking, or chewing gums in a laboratory is allowed if you are working in a clean area.

   True / False

10. All injuries, accidents, or broken/defective equipment in a laboratory must be reported immediately, even if the defect appears to be minor.

    True / False

Answers

1. False
2. True
3. True
4. False
5. True
6. False
7. False
8. True
9. False
10. True

Colour Plates

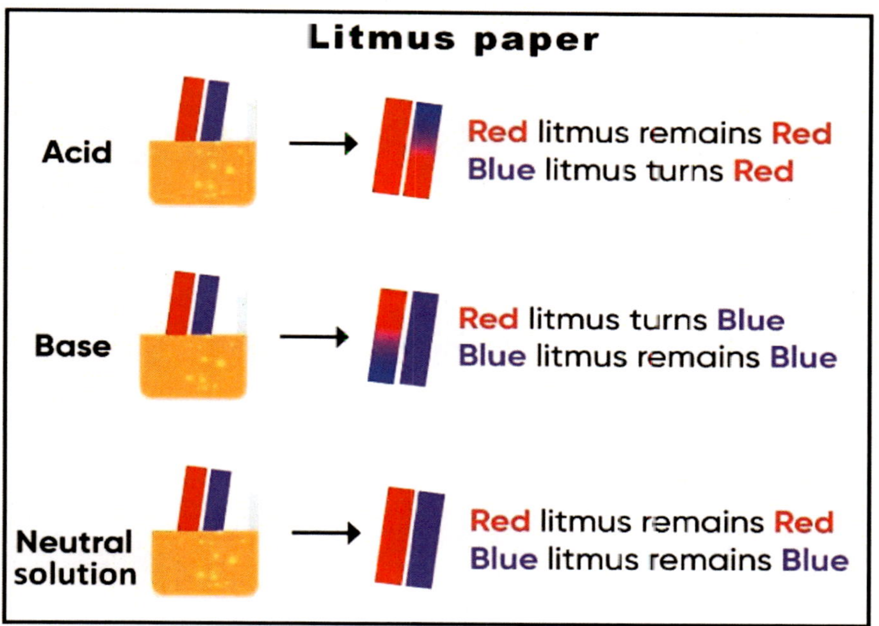

Fig. 10.3: Colour changes on litmus papers

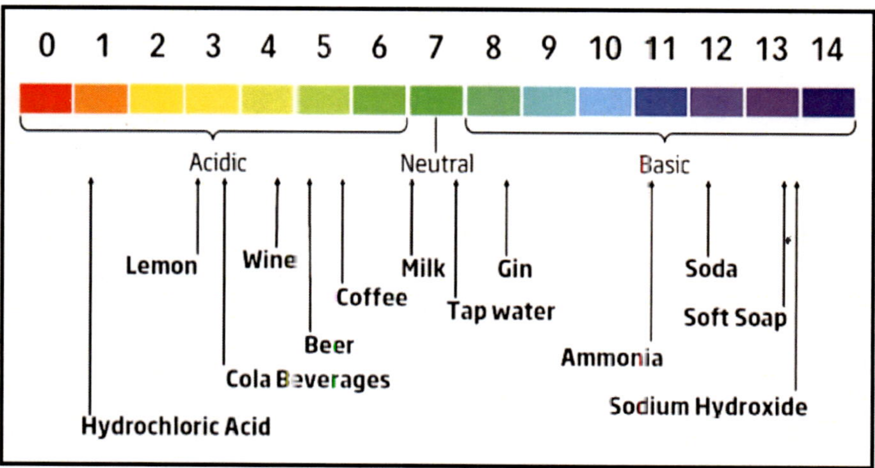

Fig. 10.4: pH of different solutions

Index